CARD

INTERNATIONAL CENTRE FOR MECHANICAL SCIENCES

COURSES AND LECTURES - No. 165

6 24.
15082
RO
1974.

Geol

ROCK MECHANICS

EDITED BY

L. MÜLLER
UNIVERSITY OF KARLSRUHE

COURSE HELD AT THE
DEPARTMENT OF MECHANICS OF SOLIDS

UDINE 1974

SPRINGER - VERLAG   WIEN - NEW YORK

ISBN 3-211-81301-2 Springer-Verlag Wien - New York
ISBN 0-387-81301-2 Springer-Verlag New York - Wien

PREFACE

Rock mechanics is one of the scientific disciplines in which progress can only be achieved by means of interdisciplinary team work. Mechanics principles play an extremely important role in the exchange of ideas and experiences amongst the fields of geology, geophysics, the science of fabrics, material science, mechanics, civil, and mining engineering.

As a branch of mechanics, rock mechanics can not prosper outside the general fundamentals of the science of mechanics, on the other hand it can not be based on the presently available theories of mechanics of continua, because discontinuities are one of the major properties of the materials rock mechanics has to deal with. What is necessary is a mechanics of discontinuous materials, which at present, however, is still in the stage of development. The need for cooperation between geoscience and mechanics seems omnipresent in this regard. Therefore we feel a great satisfaction that the young and not yet entirely mature discipline of rock mechanics has found a forum at the International Centre for Mechanical Sciences. This is the place where, out of the approach between such far apart fields as geoscience and mechanics, solutions to the major problems of the mechanics of discontinua may evolve.

I am certain that all the experts in the field of rock mechanics will be grateful and like to thank the CISM for giving them an opportunity for international and interdisciplinary work and discussion.

December, 1974

Leopold Müller

LIST OF PAPERS

L. MUELLER
"Introductory Lecture"

L. MUELLER
"Technical Parameters of Rock and Rock Masses"

L. BROILI
"Geology in Rock Mechanics"

F. RUMMEL
"A Review of Fracture Criteria of Brittle Rock"

F. RUMMEL
"Brittle Fracture of Rocks"

H.K. KUTTER
"Failure Mechanism of Jointed Rock"

R.E. GOODMAN
"Introductory Lecture of Finite Element Analysis for Jointed Rocks"

G. BARLA
"Rock Anisotropy — Theory and Laboratory Testing"

K.W. JOHN
"Engineering Properties of Jointed Rock"

N. RENGERS
"Friction Properties and Frictional Behavior of Rock Separation Planes"

H.K. KUTTER
"Analytical Methods for Rock Slope Analysis"

LIST OF AUTHORS

G. Barla	Studio Geodetico Italiano, Italy
L. Broili	Udine, Italy
P. Egger	Abteilung Felsmechanik, Universität Karlsruhe
R.E. Goodman	Department of Civil Engineering, University of California
K.W. John	Institut für Geologie, Ruhr-Universität, Bochum
H.K. Kutter	Institut für Geologie, Ruhr-Universität Bochum
C. Louis	Bureau de Recherches Géologiques et Minières, Service Géologique National, Orléans
L. Müller	Institut für Bodenmechanik und Felsmechanik, Universität Karlsruhe
N. Rengers	International Institute for Aerial Survey and Earth Sciences, (ITC), Enschede
F. Rummel	Institut für Geophysik, Ruhr-Universität Bochum

INTRODUCTORY LECTURE

Leopold MUELLER
Institut für
Bodenmechanik und Felsmechanik
Universität Karlsruhe

---

* All figures quoted in the text are at the end of the lecture.

Basic Information on Problems in Rock Mechanics

Rock mechanincs as a new field of science has grown out of the practical needs of civil and mining engineers only in very recent times and the necessity for more economical (less "overdesigned") structures in rock and aspects of safety have been the prime movers in the development of rock mechanics as a separate scientific discipline. The scope of this discipline "encompasses the engineering interpretation of geological findings; the determination of the engineering properties of in situ rock masses in terms usable for mechanics analyses and the conduct of these analyses for problems essentially related to rock masses" (John, 1962).

Table 1 briefly indicates the various subjects to be dealt with in rock mechanics, and gives the various types of engineering constructions to which principles of rock mechanics may be applied.

One of the prime reasons for the establishment of rock mechanics as a special discipline outside the older field of soil mechanics is the generally discontinuous nature of rock masses (Müller, 1961). Therefore, rock mechanics, in most cases, is concerned with the study of a jointed mass - which can only in special cases be treated as an anisotropic continuum (see Figure 1 and Table 1).

It has been one of the major contributions of the "Austrian School" to recognize the paramount importance of discontinuities in the deformation behaviour and strength of in situ rock masses and the fundamental concepts of this school have been summarized by Müller (1958, 1959, 1962):

1. For most engineering problems, the technological properties of a rock mass depend far more on the system of geological separations within the mass than on the discontinuum, that is, a jointed medium.
2. The strength of a rock mass is considered to be its residual strength that together with its anisotropy, is governed by the interlocking bond of the unit rock blocks representing the rock mass.
3. The deformability of a rock mass and its anisotropy result predominantly from the internal displacement of the unit blocks within the fabrics (structure) of rock mass.

This curiosity of our material has to enter both static and dynamic considerations concerning rock structures, which have to be built accordingly.

A further pertinent fact is that the engineer building in rock cannot choose his material but has to work with the given material of generally not accurately definable properties possessing a complex history hard to decipher in detail. The fact that any solution of rock mechanics problems requires a knowledge of the present geological situation as well as the past history of the particular rock mass, requires the collaboration of civil and mining engineers and of engineering geologists.

The spatial arrangements of discontinuities in rock masses is seldom statistically isotropic. In almost all cases, one can realize certain preferred orientations of fabric and structural elements like for example in Figure 2.

For small homogeneous areas the effect of such preferentially arranged discontinuities could be experimentally verified on both rock - and model materials. These experiments clearly indicate the resulting orientation dependence of strenght and deformation characteristics (Figure 3). For this reason the bed rock must not be considered as a bearing ground only but it has to be realized that this anisotropic deformation behaviour reinfluences the strain and stress distribution in the structure itself. Because of such intersections rock mass and engineering structure have to be regarded together as a single physical system.

In the following it will be shown, that discontinuities in rock masses may occur as bedding or ordinary joints, joints of very large dimensions, faults or fracture zones; and depending on their size, and how they influence the strength and deformation behaviour of a rock mass to various extent (see Fig. 4 and 5).

Large deformations — which are irreversible to a great extent — are a further characteristic of in situ rock masses (Fig. 6). Strain before fracture or yielding in engineering materials is generally in the range of a few promille — in compression the strain at fracture for concrete, for example, is approx. 0,4% to 0,5%. In contrast tests on block models did yield strain values up to 13% before rupture, and after the Vajont rock slide a lateral expansion of the rockmass of 16% has been determined as a result of compression of the rock mass as it reached the opposite slope of the Vajont valley.

These large strain magnitudes are a result of the movement of individual blocks along their surfaces of separation, rather than deformation of the solid material itself.

A further matter of concern in rock mechanics is the time dependence of the deformation parameters. Fig. 7, for example, examines the strain rate dependence of the deformation behaviour of slate. In situ rock masses may likely be even more sensitive to strain rate, yet little research has been undertaken on this very important aspect of the influence of time. The importance of the time factor has been realized to some extent in tunnel construction (Rabcewicz 1944 and

1962), and this has indeed led to new concepts in tunneling practice. The time factor is however equally important to surface constructions in rock masses, where it is known for example, that the rate at which excavation of the rock mass proceeds has considerable influence on the stability of slopes. Previous phases of deformation often lead to residual stresses in rock masses and in some areas stresses of unpredictable magnitude may occur from presently active tectonic deformation. As these processes are not clearly understood at the present, estimates of in situ primary stresses in rock masses are difficult to make. Nevertheless the influence of such tectonic and residual stresses is regarded as more and more important recently. These stresses – which are often characterized by horizontal stress components larger than their vertical components – are not only important in underground construction, but in my opinion also influence slope stability.

The main purpose of these few examples, the aspects of which will be discussed in a much more explicit and detailed manner in some of the following lectures, is to give an understanding of the immense complexity of the deformation and strenght characteristics of in situ rock masses and to indicate, as to why rock mechanics had to develop into a separate scientific discipline.

Difference between Soil Mechanics and Rock Mechanics Approach to the Analysis of Mechanical Behaviour

From the foregoing discussion it should be rather obvious, that the use of continuum mechanics – i.e. the soil mechanics approach – is only applicable in a limited number of special cases in rock mechanics (see for example Table 2).

Although, both soils and rock consist of assemblages of mineral grains and flakes, the mineral grains are generally cemented to a much larger degree in most rocks. But the major, more severe difference is the presence of discontinuities in rock masses which have quite decisive effects on its properties. In comparison to soils, rock masses possess a lower degree of freedom with regard to movement. The movement of the individual joint blocks generally occurs parallel to lines of intersection of various joint sets only, and the restraint against internal rotation is much larger.

Both soils and in situ rock masses are in most cases two phase systems as both generally contain water, but again, the presence of planar discontinuities which govern the flow capability, cause drastic differences between the permeability of soil and in situ rock mass and makes the in situ permeability an orientation dependant property. Moreover, the static as well as the dynamic action of the liquid on the firm phase is different in both cases; it is an isotropic pore pressure in soils, but an anisotropic effect (oriented joint water pressure) in rock masses. Some strange behaviour of rock masses must be recognised as a result of this fact, and some of them will be discussed in one of the last lectures.

Even this short discussion should have made it obvious, that the soil mechanics approach which allows directly applicable tests to be carried out on relatively small samples in the laboratory, will in most cases yield misleading results if applied to rock mechanics problems.

The points presented here, do not mean, that knowledge of rock properties and knowledge of soil mechanics is unimportant in rock mechanics, but rather, that this knowledge is not sufficient to effectively deal with the complex material called rock mass.

Different Approaches to the Problems of Discontinuous, Anisotropic Rock Masses

Although the number of scientists and engineers who by now have realized the prime importance of discontinuities of various scales in rock mechanics problems has been growing steadily around the world. But even among this group of scientists and practical engineers who do include the influence of discontinuities and the anisotropic and inhomogeneous nature of rock masses into any consideration of their mechanical behaviour, we can find adherents of two different approaches to the problem.

There is one school of thought which believes that it is possible to analytically determine the strenght and deformation behaviour or rock masses under various conditions of loading incorporating data describing the properties of the solid rock material as determined by laboratory testing of small rock specimens and the actual structural fabric of the in situ rock mass observed in the field.

An other school of scientists is convinced however, that for the present, the great complexity of rock masses is too difficult to account for, and that, in most cases, it is impossible to determine all the necessary parameters and properly account for their influence. This is a reasonable point of view for anybody who realizes the interactions between the various parameters. Once this complexity is realized, it seems logical, to determine the deformation characteristics of rock masses through large scale in situ tests and thereby obtain useful rock parameters, which can directly be used as input data for futher structural analysis in which the rock mass should always be regarded as part of the overall physical system. In my next lecture I will present my views — as a member of the second school of thought

— in more detail, while Mr. John in his lecture on the failure behaviour of jointed rock masses will present the first view point.

In such a course as ours, it seems difficult, at least in its introduction, not to repeat known facts to some degree. This has to be done to ascertain, that all participants may follow the later lectures from a comparable basis. The peculiarities of the material "rock mass" and the special problems encountered with it, make it worthwhile to recall the fundamentals of the discipline from time to time because a science like rock mechanics has to be governed by these very properties of the material rather than the present possibilities of mathematical analysis.

One may likely discover some discrepancies amongst the presentations of the various scientists during the course of these lectures. On purpose no attempt has been made in eliminating them, so as to stimulate our thinking and create an atmosphere for fruitful discussion.

The presentation of differing opinions on certain subjects should certainly not lead to frustration, but is supposed to help all participants to more clearly define their own standpoint.

REFERENCES

[1]   Bieniawski, Z.T. (1969): Deformational behaviour of fractured rock under
        multiaxial compression.
        Proc. Int. Conf. on Struct. Solid Mech. and Engg. Design. Publ. Nr.
        55 (1969).

[2]   Döring, T. (1964): Begriffsgliederung und Klassifikation der Gebirgs-
        strukturen.
        Felsmechanik und Ingenieurgeologie, Suppl. I, 1964, pp. 10-19

[3]   Döring, T. (1967): Ueber den Einfluss der Klüftung auf die Spannungs-
        verteilung im Fels.
        Felsmechanik und Ingenieurgeologie, Suppl. III, 1967, pp. 18-26.

[4]   John, K.W. (1962): An approach to rock mechanics.
        Journ. of the Soil Mech. and Found. Div. Proc. Americ. Soc. Civil
        Eng., Vol. 88.

[5]   Müller, L. (1958): Geomechanische Auswertung gefügekindlicher Daten.
        Geologie u. Bauwesen, Vol. 24, No. 1, S. 4-21.

[6]   Müller, L. (1959): The European approach to slope stability problems in
        open pit mines. Quart. Colorado School of Mines, Vol. 54, No. 3, pp.
        115-133.

[7]   Müller, L. (1962): Gestein und Gebirge.
        Sonderdruck, Festschrift "Leobener Bergmannstag", 1962.

[8]   Müller, L. (1962): Die technischen Eigenschaften des Gebirges und ihr
        Einfluss auf die Gestaltung von Felsbauwerken.
        Preprint, Gesellschaft für Boden Mechanik und Fundationstechnik,
        Luzern.

[9]   Müller, L. (1971): Die mechanischen Eigenschaften der geologischen Körper.
        Carinthia II; Sonderheft 28, Festschrift Kahler, pp. 177-191.

[10]   Müller, L., Tess, C., Fecker, E., Müller, K. (1973):
        Kriterien zur Erkennung der Bruchgefahr geklüfteter Medien — Ein
        Versuch.
        Rock Mechanics, Suppl. 2, pp. 71-92 (1973).

[11]   Rabcewicz, L.v. (1944): Gebirgsdruck und Tunnelbau.
        Springer-Verlag, Wien, 1944.

[12] Rabcewicz, L.v. (1962): Aus der Praxis des Tunnelbaues. Einige Erfahrungen über echten Gebirgsdruck.
Geologie u. Bauwesen, Vol. 27, No. 3-4.

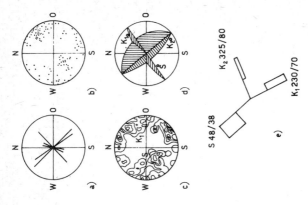

Modes of display of spatial distribution of joints: a) rose diagram, b) stereographic projection of joint poles, c) pole density diagram, d) great circle stereographic projection, e) unit square display (Müller)

Figure 2 — Spatial distribution of discontinuities in a rock mass

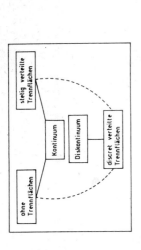

"Grundkörperkontinuum" and "Trennflächenkontinuum" as theoretical cases of the real rock mass

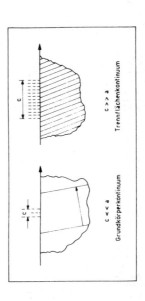

Relative attachment of a jointed rock mass as continuum

Figure 1 — Treatment of a rock mass as continuum or discontinuum

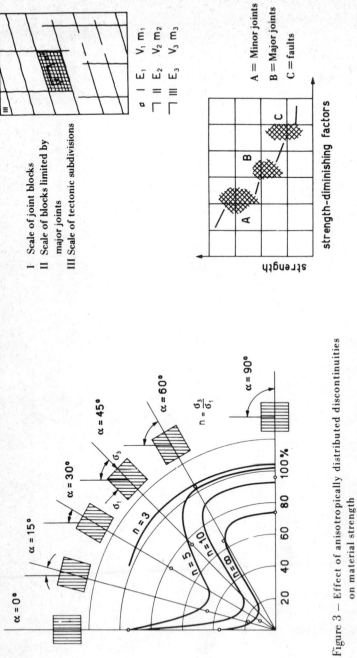

I   Scale of joint blocks
II Scale of blocks limited by
    major joints
III Scale of tectonic subdivisions

◻	I	E_1	V_1	m_1
⌐	II	E_2	V_2	m_2
⌐	III	E_3	V_3	m_3

A = Minor joints
B = Major joints
C = faults

Figure 4 —   Effect of scale discontinuities on strength behaviour of
             rock mass

Figure 3 —   Effect of anisotropically distributed discontinuities
             on material strength
             Reduction of strength as a function of angle α
             (Müller and Pacher, 1965)

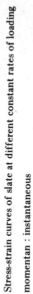

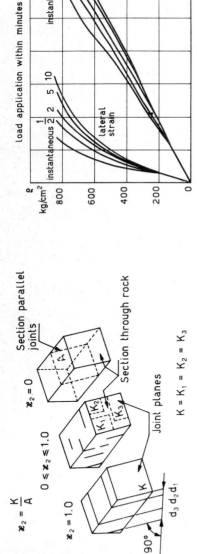

Figure 5 — Two dimensional extent of geological planes

$$\varkappa_2 = \frac{K}{A}$$

$$0 \le \varkappa_2 \le 1.0$$

$$\varkappa_2 = 0$$

Section parallel joints

Section through rock

$$\varkappa_2 = 1.0$$

Joint planes

$$K = K_1 = K_2 = K_3$$

$$K = K_1 : K_2 : K_3$$

$$90°$$

$$d_3 \ d_2 \ d_1$$

Stress-strain curves of slate at different constant rates of loading

momentan : instantaneous

Belastung in Minuten : load application within minutes
Querdrehung : lateral strain
Längszusammendrückung : axial compression
Verformung in mm : deformation in mm

load application within minutes

instantaneous $\frac{1}{2}$ 2 5 10

instantaneous $\frac{1}{2}$ 2 5 10

axial compression

lateral strain

deformation in $\frac{1}{100}$ mm

ϱ kg/cm²

800

600

400

200

0

0   5   10   15   20

Figure 7 — Strain rate dependence of deformation-behaviour
(from Döring T. 1964)

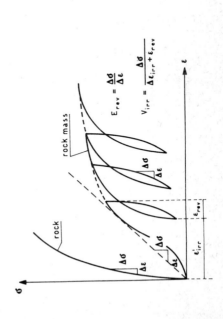

$$E_{rev} = \frac{\Delta\sigma}{\Delta\varepsilon}$$

$$v_{irr} = \frac{\Delta\sigma}{\Delta\varepsilon_{irr} + \varepsilon_{rev}}$$

rock mass

rock

$\Delta\sigma$

$\Delta\varepsilon$

ε_{rev}

ε_{irr}

σ

ε

Figure 6 — Schematical stress-strain curve for rock and rock mass

No.	type	characteristics	system regarded as	appropriate discipline	deformation behaviour D: deformation MD: mechanism of deformation CV: characteristic values B: dependent on	strength behaviour B: dependent on Str: strength SYC: strength and yield criteria CV: characteristic values	anisotropy	degree of separation
1	unfrac-tured rock	unjointed massive	single body (continuum)	mechanics of continua, rheology	B: grain fabric D: elastic, plastic, fracture, flow CV: true modulus of elasticity Modulus of Plasticity P Shear Modulus S Poisson number m toughness η relaxation time t_x	Str: true strength in tension, compression and shear SYC: fracture theories hydromechanical theory of solids CV: tensile strength compressive strength shear strength	anisotropy only due to grain fabric	$\varkappa = 0$
2	partially jointed rock mass	relatively regularly distributed micro joints	multiple or discontinuous body	geomechanics	B: transitional from No. 1 to No. 3 D: elastic and pseudo-plastic MD: fracture, cataclastic flow CV: apparent modulus of elasticity E (modulus of deformation)	B: transitional from No. 1 to No. 3 Str: residual strength SYC: presently no adequate theories available CV: residual tensile strength $\sigma_{t\,res}$ residual compressive strength $\sigma_{c\,res}$ residual shear strength τ_{res}	anisotropy caused by tectonic activity	$\varkappa > 1$
3	jointed rock mass (dissected into block-size units) jointed rock mass (dissected into small units)	abundant through-going joints	orderly arranged multiple body	structure and fabric analysis	B: type and degree of jointing, frictional properties of joints, joint-space, joint-filling D: state friction, dynamic friction MD: generally pseudo plastic cataclastic (fracture)-flow CV: inner block mobility modulus of deformation	Str: strength of rock mass SYC: presently no adequate theories available CV: app. tensile strength $\sigma_{t\,app}$ app. compr. strength $\sigma_{c\,app}$ app. shear strength τ_{app}	anisotropy caused by tectonic activity	$\varkappa = 1$
4	strongly fragmented rock mass mylonite	completely fragmented or powderized rock material	granular mass	soil mechanics fabric analysis	B: cohesion, friction, fabric, pore volume etc. D: elastic, plastic MD: differential movement of grains CV: Stiffness-number S_v permeability k water content w toughness y	Str: shear strength SYC: Coulomb criterion etc. CV: value of inner friction cohesion P_c	not yet investigated	

TABLE 1

Major subjects of rock mechanics and fields of application

ROCK MECHANICS	CONSTRUCTIONS IN ROCKS
1.1 Types of rocks and rock masses	
1.2 Basic data on the mechanical behaviour of rock	
1.3 Factors influencing the mechanical behaviour of rock masses	
1.3.1 Rock properties	
1.3.2 Fabric of rock mass (German: Gefüge)	
1.3.3 Water contained in rock mass	
1.3.4 In situ stresses	
1.3.5 Time	
1.4 Technologically important properties of rock mass	2.1 Surface constructions
1.4.1 Properties influencing strength of rock mass	2.1.1 Slopes in rock masses
1.4.2 Properties influencing deformation behaviour	2.1.2 Foundations for buildings
1.4.3 Failure mode of rock mass	2.1.3 Dam foundations
1.4.4 Properties determining the suitability for application of certain technologies of excavation, support etc.	2.2 Underground constructions
	2.2.1 Tunnels
	2.2.2 Shafts
1.5 Quantitative prediction of the mechanical behaviour of constructions in rock	2.2.3 Caverns and various large underground openings
1.5.1 Mechanical models	
1.5.2 Mathematical models	
1.5.3 Methods of mechanics of continuum and of discontinuum	
1.6 Measurement techniques in rock mechanics	

TABLE 2

Rock masses – mechanical behaviour and its relation to degree of jointing

TECHNICAL PARAMETERS OF ROCK AND ROCK MASSES

Leopold MUELLER
Institut für
Bodenmechanik und Felsmechanik
Universität Karlsruhe

---

* All figures quoted in the text are at the end of the lecture.

In the introductory lecture of this "Course on rock mechanics", the fundamental differences between the rock properties and the rock mass properties were mentioned. In this lecture, we shall examine more closely, first of all, those engineering properties of the rock masses which require to be determined in our engineering structures. It need not be emphasized that the engineering structure and the geological ground form together a physical system in which the rock or the ground is usually the weaker part, and consequently determines the overall safety of such a system. Only by a correct appraisal of the properties of the rock masses the design of an engineering structure, e.g. tunnels, slopes in open pit mines, will meet the safety requirements and will be economically dimensioned.

Even today, though the critical principles have been known for the last two decades, considerations are often based on the properties of the rock elements instead on the properties of the rock masses. In this respect it is typical that in many rock mechanics laboratories a major part of research and testing activities are still devoted to the rock material and far too little to the mechanical behaviour of the complex rock masses. How big this difference between the properties of the rock material and the geological rock body formed by this rock material can be, was very clearly illustrated by HALL (1963) in the case of the slope of a Swedish open pit mine (Fig. 1). The back calculation of the steep granite slopes gave a frictional angle of tg ρ = 0,8 and a cohesion intercept of only 3,0 kp/ cm^2 while the granite rock itself had a compressive strength, as determined from a test cube, of about 2500 kp/cm^2 and a correspondingly high shear strength.

A rock mass, i.e. a rock formation. is, in general, made up of a large number of elements. Such a material is much more complex than most other materials engineers have to deal with. It is more polymorphic, more inhomogeneous, i.e. changing from place to place. As mentioned before, the material and its mechanical behaviour is characterized and basically influenced by the two basic properties, its immense anisotropy and discontinuity.

Anisotropic with respect to its mechanical behaviour is the rock material itself which constitutes the rock mass whose strength and deformation behaviour depend on the direction. This anisotropy of the rock material is of a dimension — up to about 1:5 — which may be neglected or, by considering it as orthotropic, (or plagiotropic respectively) be treated analytically. Much greater anisotropic behaviour in regard to mechanical properties and particularly to the deformation and strength is caused by the numerous discontinuities such as cleavage and bedding planes, joints and geological faults which cut through the rock. They may be summed up here under the general term of separating surfaces or discontinuities. As a result of such separating surfaces, the shear strength in certain directions may for instance be 1/100 of that in other directions (*). Since these discontinuities are present in the rock mass in an almost unlimited number and in various directions, they can be mechanically mastered not by accounting for the influence of the individual discontinuities but only by including the properties these surfaces confer upon the rock mass into the notion or hypothetical model of a "rock mass", while distinguishing it from the idea of "rock" as a substance (CLAR, 1963).

A rock mass may be considered as an accumulation of rock elements which are bound by the discontinuities; it is not, however, a loose unorderly heap as we meet it in a slope of debris but a strictly ordered one. These existing discontinuities mostly formed as a result of tectonic overloading of the rock element. They need not always be the result of a single deformation process in the history accompanied by simultaneous rupture and flow causing great deformation; but the rock material may, under the influence of other deformation cycles, have built up new systems of discontinuities with different orientations.

STINI used to compare rock masses to well-layed dry masonry. In analogy to the soil mechanics conception of grain assemblages one might call this an arbitrary block arrangement.

Rock mass, as a rule, is a mass broken through and through due to overloading during the tectonic movement occurring during and after the period of orogenetic processes. Special importance must be attached, in this context , to the degree of joint spacing, i.e. the break-down of the material into elementary blocks. STINI specified the degree of rock discontinuity by a joint spacing index (Fig. 2a). PACHER supplemented this measurement for rock discontinuity by determining the degree of joint continuity for which purpose he defined a two-dimensional and a three-dimensional share of the joint surface (Fig. 2b). Dr. BROILI discussed the

---

(*) In the language of geologist, joints in general are frequently divided into Paraclases and Diaclases dependent on whether there was any movement of the edges parallel to the joint surface or not. For the purpose of engineering geology it is practical to omit this distinction (CLAR 1963, MUELLER 1963, STINI 1929), for any statement that a parallel movement has occurred depends upon the accuracy of observation, and it is simpler to speak of fissures (STINI 1929) or even more clearly, of joints (CLOOS 1936, SANDER 1948)!

division of material into the elementary blocks, and the joint spacing as well as the degree of joint continuity. By the degree of joint continuity, one attempts to define in how far a new fracture may follow the existing discontinuities or to which extent material bridges will first have to break down.

This dimensioning of the discontinuities must have a mechanical significance; for, if the rock mass is completely jointed by throughgoing discontinuities (e.g. box type stone arrangement or masonry stone arrangement), it could be said, in first approximation, that the principal deformation process is a "laminar slip movement" (SANDER) and that resistance against deformations is given mainly by friction along these joints. In case of a smaller degree of joint continuity, when the material is only cracked, and where the model case of the "joint blocks" is only partly achieved, material bridges will have to be sheared so that in this case two different principles may come into action (Fig. 3).

It seemed, therefore, obvious to assign a higher strength value defined in the sense of loading capacity to a material with a lower degree of jointing than to a material with penetrating joints because the shear resistance of the substance in general is higher than its frictional resistance on the joints. No attention, however, was paid when thinking in such a way, to the fact that very high stress concentrations will occur at the termination of joints due to notch action. On the other hand, the material bridges largely prevent relative shifts of the joint edges so that friction cannot be really effective on the jointed surfaces. Therefore we are not allowed to account on both the joint friction and the shear resistance of the material in the same time.

Clearly, matters are not as simple as to allow for merely using the degree of jointing as a linear reduction factor in the calculations. Still, this model of the hypothetical box type arrangement has helped us on and permitted us to find a mathematical basis which is now used in practice (BRAY, 1967; GOODMAN et al., 1968; JOHN, 1969, 1970; MALINA, 1969; WALSH and BRACE, 1966).

The strength of the rock mass is, at any rate, not the strength of the rock substance but is rather its residual strength. With stiff testing machines (COOK, 1966; BIENIAWSKI, 1966; WAVERSIK, 1968), or with servo-controlled testing machines (RUMMEL and FAIRHURST, 1970) it has beeen proved that a rock specimen does not explode violently when its maximum load bearing capacity (strength) is reached but fractures continuously with further deformation. It still maintains a certain strength, the "residual strength". This means that the stress-strain curve of a rock substance does not end with the attainment of the maximum load (upper curve, in a normal testing machine) but that, after the apperance of the decisive first fractures, the load bearing capacity of the rock drops quickly or slowly with further deformation (lower curve — Fig. 4), dependent on material and rate of loading.

We may apply in large scale to tectonic formations what we have learned in studying the post-failure behaviour of rock specimens in the stiff testing machine. Since also the tectonic forces do not act as "conservative" forces but they are comparable with the forces in a testing machine. By the influence of an unknown tectonic process during its deformation history the material had been loaded beyond its maximum load bearing capacity and intensively fractured and had become less and less stiff and resitant. Considering the geological bodies in their present state, the first part of the stress-strain curve for a certain distance beyond its peak is already a matter of the past. When testing the deformability and loading capacity of the in situ rock mass or when newly loading it, e.g. due to the construction of a project, the material will follow a stress-strain curve which is nothing else than the continuation of the curve of tectonic loading. This descending part of the curve characterizes the residual strength to be taken into account for engineering purposes. The deformation then will be associated with further fracturing and a continuous change in the mechanical properties of the rock mass, such as a decrease in the load bearing capacity and its elastic behaviour.

The residual strength of a jointed medium is guaranteed only to a certain extent by the molecular binding of the rock substance or the grain-to-grain bonding of the crystals; to a much larger extent it is the result of frictional resistance which is activated by displacements of the elements along the elements along the joints. Since initial displacement is necessary for the activation of friction, it should not come as a surprise that, when a geological body is subjected to a force, a large deformation occurs before the rock elements resist the applied forces. Not only is its module of elasticity considerably smaller (often 1/10 to 1/20) than the rock element module, but jointed material also shows, e.g. under first loading in a test, a very plastic behaviour with great compressibility and large irreversible deformations (Fig. 5). The modulus measured at first loading is commonly called the deformation modulus and set in contrast to the modulus of elasticity which can actually be known only after unloading. The relationship between the deformation modulus V of the rock and the E-modulus of the fractured rock material can be seen as a function of the degree of fracturing or jointing of the rock mass, and according to what was said above as a function characterizing the tectonic history.

From the above, it is evident that in rock mechanics., research concentrating on friction phenomena in rocks should play an important role. Very little is known as yet about these phenomena. Therefore, a special lecture on this subject will be given by Dr. RENGERS during this course.

To me it seemed worthwhile to recall, with these brief comments, the well-known, simple but essential principles of geomechanics so that the following lectures and discussions may have a common point of departure. Besides, it is useful in every scientific activity to review the position from time to time. An engineer, like an architect, must thoroughly know the material which he uses for building, and whose behaviour he wants to precalculate. For a rock engineer it is all the more important since he not only builds with the material and on it but also in it, and because his material is exceptionally complicated.

Many attempts have been made to devise handy calculation methods and simple formulas for everyday use to that the engineer may be provided with simple working rools. I see a danger in this: Complicated things do not become simpler through simplification at all cost. Things in geomechanics are complicated by their very nature.

It is true that we must aim at discovering the usually simple natural laws on which are eventually based even the most bewildering phenomena. And only when we have fully understood something in all its intricacy we can put it into simple words and formulas. We have yet to go a long way to reach that stage.

The second great danger, in my view, for the development of this science is that we are always tempted to develop what is exact, or seemingly exact, and can be handled, and to neglect or push aside those things that are not yet quite tangible, cannot yet be mastered in our formulas. But such suppressed complexes are dangerous and lead to misconceptions the inaccuracy of which escapes our minds. Looking at the literature on rock mechanics, one cannot deny that most researchers rack their brains for elegant and high-rating complex mathematical solutions while the more important basic principles and the input values of the calculations are often set aside with inappropriate leniency. One expert rightly criticized this seemingly scientific approach as "rock mechanics without rock".

This brings us to one of the most difficult problems of geomechanics, i.e., the translation of the geological data into mechanical parameters. Dr. BROILI spoke on this subject from the geological point of view, and I should like to touch upon it from the point of view of mechanics. In the introduction, we already discussed the two possible methods for arriving at useful strength and deformation parameters that can be used in subsequent calculations: The one method where one tries to include in the analysis the material property of the rock element and the geometrical conditions of the "internal shape" (SANDERS), trusting that the calculations can perform the difficult feat of accurately combining the two, as nature does. The second method which approached the possibilities of calculation more sceptically and prefers to include the mechanical interaction of material and joints in the mental model "jointed rock mass" and to test the material law of this rock mass in experiments so that finally these more complex parameters can be used as input values for a (simpler) calculation. Calculations have been carried out using these both methods.

KUZNECOW made the first calculation of the first type in 1947; model tests were conducted in 1959 in Salzburg (and were published in 1960). Both methods have been elaborated by many authors. The analytical relationships have been investigated by MUELLER and PACHER (1965), JOHN (1969), GOODMAN, TAYLOR and BRECKE (1968), MALINA (1969). The second method is presently applied by the Rock Mechanics Division at Karlsruhe and large scale in situ tests have been carried out in Japan and described by John (1961). The analytical method using the rock substance and the rock joint data to obtain the rock mass parameters will be discussed in this course by Prof. JOHN, Dr. BARLA and Prof. GOODMAN. I should like to briefly outline the other method for arriving at material laws for complex jointed media.

But first let us consider two factors which are common to both methods. They are (i) the influence of the stress field orientation and (ii) the quality of the stresses on the mechanical properties of rock.

Deformability and strength of rock masses are closely interrelated. As yet we know little about these, but tests by JOHN (1969) have given an important insight into this aspect. The post-failure curves give us further information. Apart from that, model tests (MUELLER, 1964) have shown in terms of quantity what we had earlier come to know in terms of quality, namely that loss of strength and stiffness caused by overstressing are related to the dilatation. Dr. RUMMEL and Dr. EGGER speak about this in detail.

The deformation of a rock mass is much greater than the deformability of the substance forming the rock mass while its strength is considerably lower than that of the rock substance and both vary greatly with stress orientation and nature or quality of the stress field. (By the quality of a stress field I mean the ratio (n) between the highest and lowest principal stresses, $\sigma_3 / \sigma_1 = n, \sigma_3 > \sigma_1$, compression).

A higher value of n means a closer approach to the uniaxial state of stress which is the most critical state for the rock mass, and which for example, leads in tunnels to the well-known often-observed cleavage fractures.

A lower value of n means approaching an isotropic state of stress under which the joints have little mechanical influence. The quality of the stress field as expressed by n is thus more important than the quantity value of the stresses. The purpose of rock anchoring or of the support in tunnels is exactly such an attempt to influence this value of n in a positive sense.

Surprisingly small lateral constraint ($\sigma_1 = 10\text{mP/m}^2$)   was sufficient, for example, in the Vaiont gorge rock walls, to stabilize them against 8 times overloading during the catastrophe. Just by a small force applied by anchors or through a thin gun-sprayed concrete lining the fracturing of the walls of a tunnel can be prevented. We have an analogy with the concrete pillars where circumferential reinforcement which exerts only a small force on the pillar nevertheless has a great stabilizing effect.

The orientation of the stress field has the greatest influence on the mechanical behaviour of a multi-element system. By this I mean that the angle at which the the greatest principal stress acts on the joint surface is of greater importance than its magnitude; in a purely multi-element system it is theoretically the only decisive factor. The angle under which a stress passes through a system of joint blocks is a much more essential factor for its load-carrying capacity than the magnitude of the load. If force is applied to a jointed rock mass in the same direction in which the tectonik forces causing these joints originally did act, the rock mass shows strength and lower deformation modulus; load being applied in other directions, the rock mass is stronger and stiffer. This explains why, when a dam axis is rotated through an angle of only 5-10°, a more favourable reaction may be achieved.

With greater deformations of a multi-element system a kinematic incompatibility sets in. The elements indent into each other, their edges are partially broken filling the joints with the gouge and the joints open, decreasing the friction and at places eliminating it completely (Fig. 6). The elements edge against each other and rotate (what we call external rotation) which further decreases the friction and overall resistance of the rock mass (Fig. 7); finally, new fractures occur increasing the jointing and joint continuity. Even in a jointed body with lower degree of joint continuity, the edging against and rotation occurs, especially whenever the acting stress field is different from the one that originally caused these joints. This causes movements with high stress concentration and is characterized by wedge-shaped opening of joints.

Any geometrical incompatibility means volume increase: To the volume of the solid substance the volume of the joint spaces and the volume occupied by the gouge material must be added. While a rock mass when subjected to a stress field initially should decrease in volume (MUELLER, 1948) due to the closure of the open joints and the indentation of the asperities, a volume increase is to be expected above a certain stress level (or a reduction in the rate of decrease of volume) — at least according to STINI's model of the dry masonry. This volume increase or dilatancy in the term of MENCL (by this we mean not only zonal dilatancy but the dilantancy of the whole mass) could be used as a critical value for approaching failure of the rock mass with increasing stress level.

Drawing a conclusion, this means that the increasing lateral strain is an index for the opening of the cracks, decrease of the friction and decreasing "apparent". strength; also an indication of approaching a state of dangerous loss of strength and beginning high deformation: a pointer to an imminent failure of our system.

In model tests with two joint sets conducted in Salzburg and Karlsruhe (with Mr. DECKER's collaboration) this deformation and loosening for different qualities of stresses and for different orientation to the stress field as well as for different degrees of joint continuity were examined with the purpose first of all to understand, in principle, what happens when a discontinuum is subjected to deformations. From these investigations, the following conclusion may be drawn:

Tests with a degree of joint continuity $k = 1$ (through-going joints) give a characteristic concave-shaped first section of the stress-strain curves (Fig. 8), an observation which is in accordance with the tests of KRSMANOVIC (1964) and RENGERS (1971). This shape of the curve indicates that with a small displacement, in the range of millimeters, full friction is activated. After greater displacement, the friction value decreases attaining a residual value. Similarly it is expected in a system with through-going joints that with loads which create compatible movements of the elements first an increase in the resistance against movement will take place (shown by an increase in V-modulus with increase in load) followed by an exhaustion of the resistance at the point where the stress-strain curve reaches a sharp curvature, which is something like the yield point or the plasticity limit. But rarely the material, above this limit, truly flows in a plastic manner, as is shown for example in this ($ k = 2/3$). Mostly the resistance increases further (something which we also observed in large-scale tests in the granite of Kurobe Dam) Fig. 9, though more slowly, i.e. under a smaller tangential module which points at a softening of the material. This increase is sometimes called the "hardening" but it seems to me that for our rock material a term like "stiffness limit" or "relaxation limit" is more suitable.

In an entirely different manner materials behave when the degree of joint continuity is low (k = 2/3 - Fig. 10). This material — it is quite a different one — is first stiff with high modulus and then may either suddenly break (brittle) or flow continuously. The stress-strain curves for this material are often zig-zag; above a certain definite stiffness or plasticity limit several stages of hardening and plastic flow occur subsequently because the whole mass is not involved in this process at once but part by part.

Whether a brittle (without any warning and mostly after a short total deformation) or a ductile failure (with large deformation and advance warning) occurs, does not seem to depend to much on the degree of joint continuity, but on the orientation of the stress field with regard to the joint fabric (anisotropy) — but this is not as yet fully known.

Considering next the influence of lateral stress (σ_1) which became evident in the first tests in the sixties and was confirmed by the later work by JOHN and others, it was shown that at lower values of n = σ_3/σ_1 < 5 the strength of the system is closer to the material strength than for larger values of n (n = 10) (Fig. 11). An interesting and unexpected connection was observed between the degree of joint continuity and the strength of the multi-element system. It is generally expected that the strength of a system with no through-going joints (k < 1) is higher than that a system with through-going joints (k = 1). Tests have shown that a system built of separate blocks is the strongest and (after an initial low stiffness) also has the highest modulus.

We interpret these results to the effect that a system with a low degree of joint continuity is more fracture-sensitive than a system with through-going joints as a result of the high stress concentrations at the joint tips.

The following may be said about the lateral to longitudinal strain ratio as expressed in the relationship of the ψ -lines: $\psi = \epsilon_1/\epsilon_3$ corresponds to the ratio between (total) lateral strain and longitudinal contraction and is analogous to the Poissons ratio, μ for a continuum: what is normally meant to be volume constance (μ = 0,5) in triaxial tests at plastic deformation, here in these biaxial plane-strain tests, ψ = 1 means area constance. These tests have shown, with few exceptions, that ψ indeed approaches 1,0 when the plastic or stiffness limit is approached.

This is only partially true, however, when ϵ_1 and ϵ_3 are taken in the direction of the principal stresses σ_1 and σ_3 (Fig. 12a).

This becomes even more apparent when we compare, instead, the greatest and the smallest strains and $\epsilon_{1,3}^*$ (Fig. 12b) which in the case of an anisotropic body do not correspond to the principal stress directions but rather occur at an angle of up to 25° (Fig. 13a) in our tests (this angle may change considerably during the course of the test.) How great the difference between $\epsilon_{1,3}$ and $\epsilon_{1,3}^*$ can be, is shown in Fig. 13b comparing stress-strain curves for σ_3/ϵ_3 and for σ_3/ϵ_3^*. Sometimes the maximum strain ϵ_3^* is by 75% greater than the ϵ_3 strain in the direction of σ_3.

Both the characteristic points of $\Delta F/F$ and the ψ curves correspond much more clearly to the characteristic points of the stress-strain curve when one plots these curves from the extreme deformation values $\epsilon_{1,3}^*$. The ψ^* curves, for example, indicate the plasticity limit especially clearly by reaching the value 1,0, which is not shown as obviously by the ψ curves.

Though these results were very encouraging, we were not completely satisfied since they were not as definite and as easy to interpret as it was expected. There are also many exceptions. They appeared not yet sufficiently reliable for the use as a criterion for recognizing an approaching failure in the natural rock at construction works. More reliable criteria were obtained only when we went back to basic fundamentals, namely the conception of loosening following loss of friction due to joint opening. When dilatation of joints causes loss of friction and, at the same time, according to GRIFFITH and FOEPPL, increases stress concentrations, it seemed worthwhile to investigate, as possible criteria, those deformations, which indicate most clearly joint opening, i.e. deformations normal to joints, and displacement along the joints. And, indeed, these deformations ϵ_A and ϵ_B normal to the joints sets KK_A and KK_B (Fig. 14) indicate such clear relationships that the figures speak for themselves.

It is clear from the above that in describing the properties of rock masses, one cannot do entirely without costly in situ tests. In my opinion, it is only their results that permit us to correctly judge the strength and the deformation behaviour of the rock mass. Such large-scale tests were, for the first time, conducted by the INTERFELS, Salzburg, in Japan and described by JOHN (1961) and NOSE (1964); (Fig. 15). These tests were conducted on rock blocks up to 16 m^3 subjected to shear and triaxial tests (not truly triaxial).

To start with the most basic things, these tests proved that the rock mass strength is much smaller than the rock substance strength and many of us were surprised to see how much lower it is. The reduction of the rock mass strength in relation to the strength of the rock substance varies from rock to rock and even in the same rock depends on the spatial attitude of joints. It may even occur that the rock mass strength is nearly as high as the rock element strength. The reduction even varies with different directions.

That the deformation and the elasticity modulus of the jointed rock mass is smaller than that of any rock element has been known for a long time and was confirmed not only in these tests but even earlier by numerous seismic and ultrasonic measurements.

Contrary to other types of tests, it was possible in these and other large-scale in situ tests to achieve an approximately homogeneous state of stress so that rock mass moduli for a certain load value in relation to, and dependence on, the stress field could be obtained.

Some in situ rock tests show a much larger permanent deformation than expected. They are often up to 50 % of the total deformation. This permanent deformation occurs not only at higher stress levels but already during initial loading, a fact, which is rather surprising. Fig. 9 shows clearly the difference between the deformation modulus and the elasticity modulus on unloading from quite small stress levels.

Fig. 9 shows that there occurs no real failure in the rock mass. One must, therefore, speak not about the failure strength, but about the "failure flow limit". A true fracture as it is observed in the laboratory tests of rock specimens and accompanied by loud noise, does not occur in the case of a jointed rock mass because this mass has already been tectonically broken and because load in the rock mass usually corresponds to a stiff load system. As such, changes in the load play a more important part in the case of a rock mass than for an unbroken test specimen.

Rock mass (on the other hand) is often more sensitive to repeated load changes than the unbroken rock element — another technical difference between rock mass and rock element. In the case of dams, the strongly changing loads due to changes in the water level in the reservoir that act upon the rock play an important role.

It was, therefore, clear that technological tests in rock mass must be long-term tests. This is also true because a rock mass shows a great elastic after-effect and a considerable relaxation — as geologists have known for a long time; it behaves like a visco-elastic liquid.

The number of load changes which, in our tests in Japan, amounted sometimes to more than 100 for each test body, required testing times ranging from a number of days to several weeks. Since the geological bodies do not show any fracture limit, but only a more rapid and intensive flow with higher loads, we had to define certain limiting strains in order to characterize their behaviour in carrying load. In many tests the limiting stress is easily visible which, when exceeded, results in accelerated deformation. These values were at the time called flow limits I, II etc., and later on stiffness limits (Fig. 9). Another possibility of relating the rock mass deformation behaviour to the strength behaviour is to record the irreversible deformation in dependence on the reversible deformation. With increased and with repeated loading the quotient of the two increases.

Calculations and model tests must be checked-up on by in situ measurements. These should not be limited to the period during which the geomechanical properties of the rock are determined before construction of the project but they must also be an important and integral part of the supervision of the project and the surrounding rock during construction time and after its completion. They permit us to check on the assumptions and calculations made and guarantee the existing safety of the structure.

Comparisons between calculated and measured stresses as are the rule in other branches of engineering are not yet possible for rock projects due to the difficulties in the measurement of stresses, safety specifications and determination of the material strength. It is somewhat easier to compare the calculated and measured deformations, but the calculation of deformations in jointed bodies is difficult even today. In such a situation one welcomes the possibility of giving criteria for the approach of failure or flow conditions alone on the basis of measured deformations, i.e. on the basis of deformation values which need not be related to any calculated values, to strength, material constants, or expected deformations but only need a comparison between measurements as demonstrated above. Deformation measurements are simple; they are nowadays carried out in every dam project, in every tunnel. One should, however, know, where and what is measured. We still have a lot to learn.

REFERENCES

[1]   Bieniawski, Z.T.: Mechanism of rock fracture in compression, Rep. Counc. Scient. Ind. Res. S. Afr. MEG 459, 1966.

[2]   Bray, J.W.: A study of jointed and fractured rock, Teil I and II, Felsmech. u. Ing. Geol., Vol. V/2-3, 1967, S. 117-136 und V/4, 1967, S. 197-216.

[3]   Clar, E.: Gefüge und Verhalten von Felskörpern in geologischer Sicht, Felsmech. u. Ing. Geolo., VOl. I/1, 1963.

[4]   Cloos, H.: Einführung in die Geologie, Bornträger, Berlin 1936.

[5]   Cook, N.G.W. und J.P.M. Hojem: A rigid 50-ton compression and tension machine, South African Mec. Eng. 16, p. 89-92, 1966.

[6]   Goodman, R.E., R.L. Taylor und T.L. Brekke: A model for the mechanics of jointed rock, J. Soil Mech. and Found. Div., Proc. ASCE, Vol. 94/SM3, May 1968, p. 637-659.

[7]   Hall, B.: The observed strength of a rock mass, Proc. Int. Conf. Soil. Mech., Budapest 1963.

[8]   John, K.W.: An approach to rock mechanics, J. Soil Mech. and Found. Div., Proc. ASCE, Vol. 88/SM 4, 1962, S. 1-30.

[9]   John, K.W.: Festigkeit und Verformbarkeit von druckfesten, regelmässig gefügten Diskontinuen, Diss. Arbeit, Universität Karlsruhe, 1969.

[10]   Krsmanovic, D.: Initial and residual shear strength of hard rocks, Géotechnique 17, S. 145-160, 1967.

[11]   Kuznecov, G.N.: Mechanische Eigenschaften von Gestein, Moskau 1947.

[12]   Malina, H.: Berechnung von Spannungsumlagerungen in Fels und Boden mit Hilfe der Elementenmethode, Diss. Arbeit, Universität Karlsruhe, 1969.

[13]   Müller, L.: Von den Unterschieden geologischer und technischer Beanspruchungen. Geomechanische Probleme I, Geol. u. Bauwesen, 16, H. 314, 1948, S. 106-161.

[14]   Müller, L. und F. Pacher: Geomechanische Auswertung geologischer Aufnahmen auf der Schachanlage Neumühl, Unveröff. Ber. 1956.

[15]   Müller, L.: Der Felsbau I, Enke Verlag, Stuttgart, 1963.

[16]   Müller, L.: Die technischen Eigenschaften des Gebirges und ihr Einfluss auf die Gestaltung von Felsbauwerken, Schweiz. Bauz. 81, H. 9, S. 125-133, 1963.

[17] Müller, L.: Beeinflussung der Gebirgsfestigkeit durch Sprengarbeiten, Felsmech. u. Ing. Geol., Suppl. I, 1964, S. 162-177.

[18] Müller, L. und F. Pacher: Modellversuche zur Klärung der Bruchgefahr geklüfteter Medien, Felsmech. u. Ing. Geol., Suppl. II, 1965, S. 7-24.

[19] Müller, L. et al.: Kriterien zur Erkennung der Bruchgefahr geklüfterer Medien — Ein Versuch, Rock Mechanics, Suppl. II, 1973, S. 71-92.

[20] Pacher, F.: Kennziffern des Flächengefüges, Geol. u. Bauwesen 24, H. 3/4, 1959, S. 223-227.

[21] Rengers, N.: Unebenheit und Reibungswiderstand von Gesteinstrennflächen, Diss. Arbeit, Universität Karlsruhe, 1971.

[22] Sander, B.: Einführung in die Gefügekunde der geologischen Körper, Springer, Wien, 1948.

[23] Stini, J.: Technische Gesteinskunde, 2. Auflage, Springer Verlag, Wien, 1929.

[24] Walsh, J.B. und W.F. Brace: Elasticity of rock: A review of some recent theoretical studies, Felsmech. und Ing. Geol., Vol. IV/4, 1966, S. 283-297.

[25] Wawersik, W.R.: Detailed analysis of rock failure in laboratory compression tests, Ph. D. Thesis, University Minnesota, 1968.

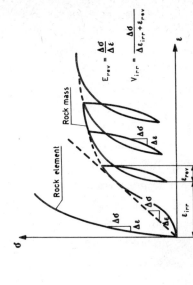

Fig. 4: Stress strain curve obtained in
a) stiff machine; b) normal machine

$$E_{rev} = \frac{\Delta\sigma}{\Delta\epsilon}$$

$$V_{irr} = \frac{\Delta\sigma}{\Delta\epsilon_{irr} + \epsilon_{rev}}$$

Fig. 5 : Schematic representation of stress-strain curves for rock
element and rock mass

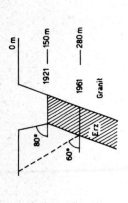

Fig. 1: Natural slope in a swedish open pit mine

Fig. 2

a) Joint spacing after STINI; k = number of joints/meter

b) Linear degree of jointing $\varkappa = \frac{\sum l_i}{l} \leq 1$

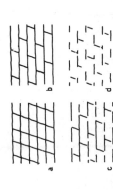

Fig. 3 : Types of jointed rock mass

a) box-type stone arrangement; b) masonry type arrangement
c) staggered arrangement; d) connected arrangement

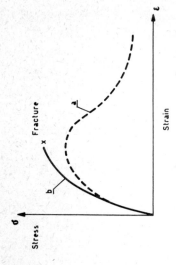

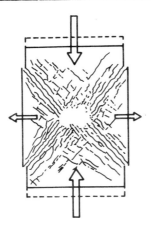

Fig. 6: Rock loosening on failure

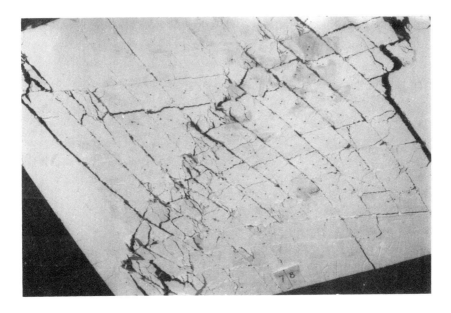

Fig. 7: Hindered rotation of single blocks and the block system of the jointed body during
the test

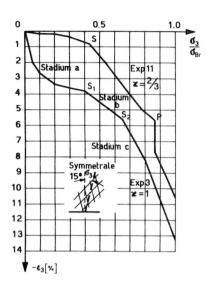

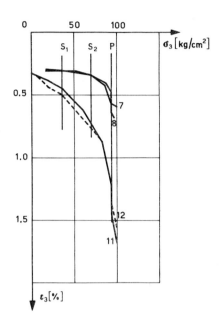

Fig. 8: Stress-strain curve for the test no. 3
(n=10, $\varkappa$=1.0) and no. 11 (n=10, $\varkappa$=2/3);
S stiffness limit; P plasticity limit, σ_{Br} highest
value of σ_3 sustained by the model

Fig. 9: Stress-strain curves from the large
scale triaxial test at Kurobe IV dam; symbols as in fig. 8

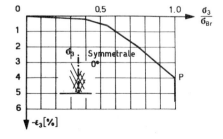

Fig. 10: Stress-strain curve for test no. 9
(n = 10, $\varkappa$ = 2/3); symbols as in fig, 8

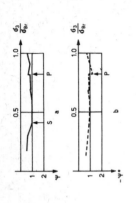

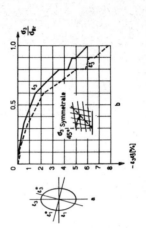

Fig. 12: a) Ratio of (lateral/longitudinal) deformation Ψ test no.10 ($n = 5$, $\varkappa = 2/3$)

b) Ratio of (maximal lateral/maximal longitudinal) deformation Ψ^* at the same test. Symbols as in fig. 8

Fig. 13: a) ϵ and ϵ^* in polarcoordinats; b) stress-strain curve for the test no. 16 ($n = 5$, $\varkappa = 2/3$)

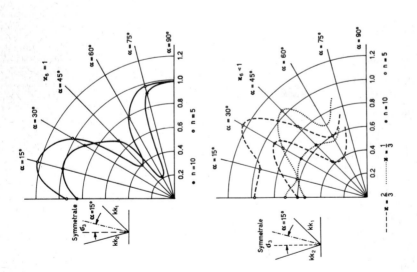

Fig. 11: Decrease of strength as a function of the angle a
a) degree of jointing $\varkappa = 1.0$; $n = 10$ and $a = 5$;
b) degree of jointing $\varkappa = 1.0$; $n = 10$ and $n = 5$

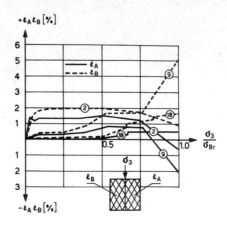

Fig. 14: Strain normal to the joints for the test no. 2 ($\alpha = 0^0$, n = 10, $\varkappa = 1.0$), no. 9 ($\alpha = 0^0$, n = 10, $\varkappa = 2/3$) and no. 18 ($\alpha = 0^0$, n = 10, $\varkappa = 1/3$)

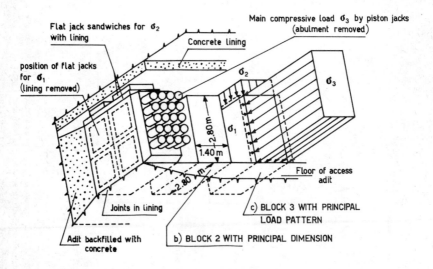

Fig. 15: In-situ triaxial compression tests on rock mass

GEOLOGY IN ROCK MECHANICS

L. BROILI
Udine — Italy

---

* All figures quoted in the text are at the end of the lecture.

First of all I wish to thank the Authorities of the International Center of Mechanical Sciences, and in particular Prof. Olzak, Prof. Sobrero and Prof. Müller, who invited me to deal with a subject which is somehow a premise for understanding the substantial reasons that suggested and favoured the development of a Mechanic of rock masses, that is a Mechanic of a geological medium.

This lecture may be entitled "Geology in Rock Mechanics", because, in the confined time at disposal, it aims to emphasize the decisive contribution of Geological Sciences to Rock Mechanics, as background and as an actual means for a correct development of a geomechanical research.

The **material**, whose mechanical behaviour we are going to investigate, under the influence of natural fields of stresses, or under artificially imposed sollicitations, is, by definition, a geological material: the "**rock**", or better, the "rock mass".

Therefore, since the **material** is a geological medium, it seems natural that Geological Sciences have to provide their own basic research methods to collect and to describe the parameters governing the mechanical behaviour of a rock mass.

Among these Geological Sciences, we can mention in particular Stratigraphic Geology, Structural Geology, Mineralogy, Mineralography, Petrography and Geophysics.

The treatment of this subject, i.e. Geology in Rock Mechanics, aims both to emphasize the mentioned determinant place of Geology in Rock Mechanics, according to the fundamental principles upholded by the Salzburg School of Rock Mechanics, and to submit autonomous elements for judging how correct and acceptable some current tendencies are which mistake "Rock Mechanics" with a Mechanics of a "material" which, on the contrary, according to the scale problem, may be only the matrix of the "rock mass". These studies represent a wasting of great scientific efforts, because they produce mathematical, physical and technological results, which by no means correspond to the mechanical reality of the given problem.

In particular in this lecture the following subjects will be treated:
- 1st Discussion on the definition of "rock mass", as referred to concepts of inhomogeneity, discontinuity and anisotropy closely related to the scale (i.e. the dimensions) of a given geomechanical problem.
- 2nd Considerations on the "**rock material**" that is the "rock" visualized like matrix of the solid phase which builds up the rock system.

— 3rd Considerations on the geological structures, that is on the geological structural elements which break off the continuity of the rock matrix, and which intervene as fundamental elements characterizing the physical system, to which rock mass can be referred.

— 4th Considerations on the extrapolation of geological data and on the rock mass classification.

Dealing with the above, quoted subjects, the basic geological elements are actually presented, which will be mentioned and discussed in the following lectures. In particular the concrete current possibilities of numerical quantification of geological parameters will be emphasized.

To begin a further reference may be necessary to the fundamental laws of Rock Mechanics, as already mentioned in Prof. Müller's premise.

The first axiom, supported by the Salzburg School of Rock Mechanics establishes that the mechanical properties of a rock mass are based more on the mechanical properties of plane structures (geological discontinuities) than on the properties of the rock considered as "rock material". Therefore Rock Mechanics is essentially Mechanics of the structures and, respectively, Mechanics of a discontinuous.

The second axiom establishes that the strength of a rock mass is a residual one, of a more or less anisotropic medium (system) (VERBAND) built up by a multitude of rock bodies having differing bond degrees (i.e. differing bond strenghts).

The third axiom establishes that the deformation properties of a rock mass depend more from the partial mobility of each single element (body) forming the rock mass, that from the intimate deformability of those elements due to the geotechnical characteristics of the material.

The fourth axiom establishes that both mechanical strength and deformability, and tension distribution within the rock mass, depend from the structural characteristics of the rock system.

Finally, the fifth axiom establishes that the anisotropy in the geomechanical properties of a separated (fractured) rock mass can be derived with statistical validity from the morphological anisotropy of the plane structure intersecting the rock mass, and that the technological properties of such a mass can be determined only if the technological tests are applied to rock masses reproducing the natural structural conditions in a statistical context. It may be added that the geomechanical properties, and therefore the mechanical response of a rock mass to

applied stresses, can result substantially modified or conditioned by pre-existing
states of virgin geological stresses, which must be for that reason determined and
carefully considered in the analysis.

Moreover **water**, which represents the liquid phase within the **rock
system** (the solid phase), may be a determining factor in modifying the physical
state and the geomechanical behaviour of the rock mass.

Thus we have generally introduced the concept of **"rock mass"** instead
of the concept of "rock". Moreover we have mentioned notions like "rock
material", "plane structures" (geological discontinuities), "latent virgin state of
stresses". Finally we have pointed out that hynomogeneity, discontinuity and
anisotropy are characterizing factors for the "medium" or for the "material" i.e.
the "rock mass" in a typical geomechanical problem.

However, to detect completely and correctly this material, that is the
"rock mass", may be difficult, because the concept is not absolute and perhaps
somehow changeable, being closely related with the scale (i.e. dimensions) of a given
geomechanical problem.

Therefore it is important, in defining what is the rock mass, to
introduce the concept that the definition of "rock mass" must be a relative
definition.

As we know, in a very high number of cases there are no distinct
boundaries between the operating field of Rock Mechanics and the one of Soil
Mechanics.

Anisotropy and discontinuity conditions may be actually present also
within the so called **soil material**. Let us consider for instance, some lithoid
heterogeneous imbedded sequences or some densely lenticular rock complexes,
which may correspond to rock masses containing weathered inclusions.

There is no substantial geotechnical difference, in these circumstances,
between an horizontally stratified sequence in the whole made of consolidated clays
and of slightly bounded sands, and a sequence horizontally stratified made up of a
sandstone completely separated by close and continuous joints, that is a completely
loose lapideous number, and of beds of soft marl.

Both situations are schematically represented in the projected slide.

The same considerations could be extended to the cases, where
conditions of complete continuity and isotropy of the material are met. In such
circumstances it is less important to distinguish whether the considered "medium" is
a lithoid homogeneous material (Rock Mechanics) or an homogeneous continuous

soil material (Soil Mechanics), the difference being only upon the actual specific mechanical strength which can be produced in both quoted different circumstances.

Likewise, when conditions of total structural separation are displayed, there is no more substantial difference compared with geomechanical behaviour, between an incoherent soil (for instance sand or gravel) and a rock mass transformed by intensive jointing or by crushing processes into a loose volume of small, completely separated unit rock blocks.

Referring to the mentioned considerations, we understand that if compared with the mechanical reponse of a geological medium, we may correctly speak about "Rock Mechanics" only when the basic conditions emphasized by the before mentioned axioms are completely accomplished.

Moreover, we can introduce the concept of **scale factor** because it is by now evident that by changing dimensions of the considered rock volume, also the inhomogeneity, discontinuity and anisotropy conditions can substantially change in parallel lines. On that account the geotechnical problem too can assume a different aspect and can require completely diverse solutions.

Therefore, the "**scale factor**" is a determining parameter of geometrical nature which provides that the actual "**dimensions**" and characters of the geolithological complex are related to its potential mechanical reply compared with applied stresses.

These "**dimensions**" vary from the dimensions of the rock crystal to the ones of a tectonic region which can include entire portions of terrestrial surface.

Moreover they are comprehended between the dimensions of artificially applied stresses and the ones which are displayed by the nature during an orogenic process under particualr conditions of temperature and time.

Actually, we usually reduce the whole problem to dimensions more close to the limits of the scientific research and to the application fields pertinent to the human operating practice.

Then the "**rock mass**" assume the dimensions of the scale to which it appears in engineering and mining works, and so it might be regarded as the study of the behaviour and properties of accessible rock masses under stress or change conditions.

In spite of being useful to the praxis, this reasoning is not always correct, in particular because at this dimension the scale problem, referred to the geological scale, to the applied sollicitation scale and to the scale of boundary conditions, may not be resolved at all, that means it remains and it still affects the geological-geomechanical foundation of the problem and, unfortunately, the results of the study. On account of this, it is basic to refer in all circumstances the foundation of the study and its results to the overall scale of the investigated rock mass. (see Fig. 1).

Therefore the scale of the geomechanical problem corresponds to the basic parameter which governs the first phase of the geomechanical study, that is the geological description and the geological classification of rock material and of rock structures, and the determination of hydraulic details and the eventual states of virgin stresses.

Somehow conditioned by the scale of the problem are also our considerations on "homogeneity zones". The definition of these zones, which are homogeneous if compared with parameters and conditions, is closely related with the concept of resolution, extrapolation and interpretation of geological, hydraulic and geomechanical parameters.

It is important to emphasize that conditions of inhomogeneity, discontinuity and anisotropy may be met already at the scale of the crystal and at the scale of an association of crystals, that is at the scale of that rock material which is currently considered as the mechanically isotrope "matrix" of the rock mass.

Thus, if the scale of our problem has dimensions comparable with those of crystal associations, then our "matrix" must be necessarily searched at the dimensions of the crystal grid.

THE ROCK MATERIAL

"Rock", or better "rock material" is the "matrix" of the solid phase of the rock mass.

John in 1962 in his "Approach to Rock Mechanics" defines the rock material and its pertinent properties as the "material" not subjected to geo-mechanical anisotropies (jointing), for instance drill cores or hand samples.

This definition is definitely right and particularly significant when we refer to the scale of the usual geomechanical problems, and in particular to the scale of engineering problem.

It may be no more significant when the scale of the problem is reduced below the dimensions of hand samples, or is increased to dimensions of tectonic regions.

Jaeger, in his book "Fundamentals of Rock Mechanics", dealing with the mechanical nature of the "rock", writes:

— "most rocks comprise of aggregate of crystals and amorphous particles jointed by varying amounts of cementitious materials. The chemical composition of the crystals may be relatively homogeneous, as in some limestones, or very heterogeneous as in granite. Likewise, the size of the crystals may be uniform or variable, but they generally have dimensions of the order of inches and small fractions thereof. These crystals generally represent the smallest scale on which the mechanical properties are studied.

On the one hand the boundaries between crystals represent weakness in the structure of the rock, which can otherwise be regarded as continuous. On the other hand the deformation of the crystals themselves provides interesting evidence concerning the deformations to which the rock has been subjected".

Jet rock material is formed by aggregates of idiomorphic crystals (that is first-generated and completely developed in each face), and of allotriomorphic crystals (that is successively formed and adapted to the former ones).

These aggregates, from a crystallographic point of view, can be irregular, that is without relations of spatial orientation like, for instance, almost all vulcanic extrusive rocks and some sedimentary rocks. Eruptive rocks, in particular, can present **"granular aggregates"**, that is generated from equidimensional crystals; in those aggregates it is possible to recognize the direction of the most important crystallographic elements. So we can distinguish a "fabric" or, better, a genetic microfabric, already at the crystal scale. Such structure, as we have already said, is not without significance for the study of the mechanical behaviour of the rock material at determined dimensional levels.

Particularly interesting are the crystalline structured characteristics in the methamorphic rocks, which permit fundamental deductions and extrapolations on the genetic processes and on the deformation processes which developed within entire tectonic regions of the earth crust.

The methodology of microstructural analysis in the ambit of crystal grid has anticipated the researches on the geological macrostructures, and has provided to these investigations both methods of data statistical collection, and the means of their representation and elaboration (Sander). (see Fig. 2).

This figure shows, on equivalent projection, (Schmidt net) the microstructural diagram of crystalline structures determined by means of optical method with polarization microscope.

We see the characteristic distribution of the "Z" axis of 400 turmaline crystals and, below the orientation diagrams of muscovite (on the left) and of the quartz. (on the right).

In particular it has to be observed that, already at a crystalline stage, some preliminary considerations may be done on the preferential orientations of geological microelements and on the concepts of structural anisotropy (statistic anisotropy) of crystalline elements.

However this structural anisotropy, at crystalline dimensions, is not only concerned with the genetic process, that is with the intercourse among crystals in correspondence with their boundaries and orientation of their symmetry elements.

Jet crystals are continuously subjected to surrounding influences and those influences necessarily cause variations and deformations in the crystal itself and in the crystalline grid.

Temperature and pressure are the factors which can, above all, deform the grid and directly intervene on the crystal. Particularly with regard to the pressures, they can be hydrostatic ones. In this case these pressures generally cause reductions of the total volume of grids: under an homogeneous and isotropic field of stresses, from spherical symmetry conditions, we generally pass to elipsoidic symmetry, because the contraction of lengths under the influence of an hydrostatical pressure may develop in a different way, according to different directions.

This observation shows already an anisotropic behaviour at the grid level.

If stresses are compressive ones, or tensile stresses acting towards only one direction, a particular distribution of tension displays within the crystal grid, acting in several directions and producing mutual pluridirectional slipping. Against these stresses the crystal grid reacts differently according to the direction of applied sollicitations, and modifies its shape in the ambit of elastic deformations or of plastic deformations when the strengths which tie the structural crystal grid, have definitvely been overcome.

The mechanical influence of a stress one-sided applied to an isotropic body, is difficult to describe and to handle, because of the great number of variables intervening in the process.

Moreover, if we treat forces acting on complex crystal structures (crystal associations   rock material), the resulting difficulties are greatly increased, compared with the anisotropy of such structures.

In this reasoning are contained already some anticipations on the considerations which will be derived later on when we shall speak about the **rock mass** for increasing scales. In particular, with reference to the crystal grid, we can observe that not only the distribution of internal tensions produced by stress application has to be considered, but also attention must be paid to the distribution of forces acting among the different particles along the different planes and towards different directions.

Within the infinite particle lines and within the infinite rational planes, which may be traced into a spatial grid, the most different stability conditions must be present and therefore the most different behaviour compared with normal forces producing bodies separation from the grid, and to shear forces producing slidings.

Plastic deformations of the grid may be **both** rupture and separation because strengths acting among particles have been overcome, **and** angular deformations producing bending of particle lines and of particle planes.

In addition, slippings with translation of a part of the grid may also happen. These slippings can be considered as the most common deformation means within the grid.

In the crystal grid, formed by one kind only of particles, the rational planes of potential sliding are generally disposed in parallel lines with the planes in which the density of the particles is the most elevated.

Normal forces finally produce fracturing and flaking conditions of crystals.

The fracture process displays mostly along uneven planes, which cannot be referred to actually defined crystallization planes.

Generally the microfracture surface is normal to any direction of least cohesion, when within the quoted plane, directions of maximum cohesion are apparent.

The diversity in cohesion generally depends from major or minor distance actually existing among particles of the crystal grid. Moreover the rupture process involves always the least bond's number.

As we have already seen at crystal grid levels a marked microstructural anisotropy exists. This anisotropy can be exactly quantified by means of a microstructural, symmetrological, statistical analysis. This anisotropy in its most apparent forms might sometimes still have significance when the scale of the problem increases until dimensions by which we usually neglect the crystal grid anisotropy , considering the crystal associations — the "rock material" — as homogeneous and continuous: that is at the dimensions of macrostructures, i.e. geological macrodiscontinuities.

The mentioned situation is, in particular, important when we refer to metamorphic rocks, by which the microsturctural anisotropy may still have significance at **rock material** levels in determining the mechanical response of the **rock mass.**

At the dimensions of the **rock material** a further cause of anisotropy can be searched simply in the particular mineralogical constitution of the rock material and in the physical and chemical modifications of the lithotype.

That means, a rock material has different properties because sometimes it may be built up of associations of crystals having different nature and properties or the mineralogical nature of these crystal associations may have been transformed by weathering or by similar weakening or modifying processes.

In these cases it is evident that different rock materials and their alteration products have inherently different weakness and strengths resulting from their formation and subsequent history.

Most important of the rock properties is the nature of mineral assemblage and the strength of the constituent minerals. A rock material cannot be strong if its mineral constituents are weak. On the other hand, if the minerals are strong, the rock may still be weak if it has unfavourable microfabric, as we have already seen. A rock can only be as strong as its weakest part. Rock materials containing soluble minerals such as rock salt, gypsum, limestone and dolomite, are particularly susceptible to dissolution and to physical alteration particularly with changes of moisture content, which can lead to extensive and rapid breakdown of rock materials due to slaking action.

Hydrothermal waters, aggressive waters, temperature conditions produce on the other hand extensive weathering of crystalline lithotypes as granite, diorite, basalts, gneiss, and so on.

Rock material can be completely described from a geological point of view, referring to well-known methods of mineralogy and of petrography. By means of these methods we establish the mineralogic, crystallographic and chemical characters of the material.

From a geomechanical point of view the strength and deformability properties of the **rock material** might be satisfactorily determined in laboratory on samples.

The dimensions of rock samples and the experimental methodology must always be representative of the scale of the investigated problem, in order to accomplish tests suitable to be correctly interpreted.

The most usual technological tests for the rock material are the following: (see Fig. 3)

1) uniaxial compression test
2) uniaxial tension
3) triaxial compression
4) extension in triaxial test

5) tests with homogeneous polyaxial stress
6) Hallow cylinder with axial load
7) diametral compression (brazilian test).

THE ROCK MASS

Having described the rock material, let us now consider the rock mass, increasing progressively the dimensions of the rock volume concerned in a given geomechanical problem. With increasing dimensions the geomechanical significance of microdiscontinuity and of microanisotropy conditions of crystal associations which form the rock material, decrease progressively and the influence of the geological macrostructure, which is typical of the **rock mass**, progressively emphasizes itself.

Thus by increasing of the scale of the problem, microdiscontinuity and then microanisotropy may be included in considerations concerning the rock material which is regarded now as the matrix of the rock system.

Compared with the new scale of the problem, the **matrix** may be mostly considered as an homogeneous, continuous and isotropic medium.

Therefore we must distinguish between the matrix (rock material) and the rock mass, which at its scale includes all **elements** producing macrodiscontinuity, macroinhomogeneity and macroanisotropy.

These elements are:
— states of lithological overall discontinuity and inhomogeneity
— conditions of structural discontinuity and inhomogeneity
— inhomogeneous distribution of water or of rock hydraulic parameters
— inhomogeneous presence and discontinuous distribution of virgin tensions within the rock system.

At dimensions including all these parameters, which are certainly the most usual in our problems, being the geological structures (geological discontinuities) a major influencing factor among the before quoted elements, the first **axiom** of Rock Mechanics assumes particular significance: that is, the mechanical properties of a rock mass are governed, above all, by the mechanical and hydraulical properties of plane structures.

Virgin tension states are also somehow connected with plane structures when we refer to tectonic features and to tectonic processes.

The mentioned axiom is therefore of paramount importance because this preposition defines in the most complete sense the essence of Rock Mechanics, that is Mechanics of a **geological discontinuum**.

On that account it is important that the geological elements produce the particular conditions we have described, be recognized, described and above all quantified, and then introduced into the analysis of the geomechanical problem as determining parameters.

Moreover it is important to clearly fix the boundaries of the rock space within which the performed descriptions and quantification of geomechanical parameters can be applied and extrapolated with statistical validity.

Let us consider in some detail the elements which produce conditions of mechanical anisotropy.

At the scale of the rock mass, discontinuity states of simple lithological order may exist. These discontinuities are especially evident in sedimentary rock complexes in which, because of horizontal or vertical face passages, gradual or sudden variations may happen within the mineralogical constitution of the rock matrix.

Differing crystalline associations can produce variations in strength properties and in mechanical deformability which, at the scale of the rock mass, finally produces a real different mechanical behaviour of the rock system.

Sometimes, compared with some particular loading directions. Regular discontinuities of lithological order may assume eminent importance, greater than structural discontinuities.

In metamorphic rock complexes, lithological discontinuities can be related to contact zones, or to allogenic intrusions or to veins of mineral or to alteration and decomposition processes.

The same thing is the case of intrusive and effusive crystalline rock complexes by which lithological variations are mostly due to contacts of different crystalline associations or to the decomposition of lithotypes because of hydro-thermal decompositions or of fault phenomena (mylonites).

Lithological discontinuities are recognized in the field and described both in the field and in laboratory, according to the usual methods furnished by the geological survey (direct or indirect, i.e. borings and geophysics), and by mineralogical-petrological methods.

The mechanical quantification of the incidence of lithological discontinuities compared with the overall anisotropy or with the orientated anisotropy, can be performed by technological "in situ" determinations, especially by means of geophysical survey (determination of dynamic modulus) of elasticity.

We may now speak about discontinuities of structural order. The description and quantification of **geological-structural discontinuities**, as we have seen, is very significant in geomechanical investigations.

The most peculiar character of a rock mass, which is of immediate evidence even to persons not accustomed to geological or geomechanical problems, are the separation conditions produced by bedding joints and by other joints belonging to more or less spreading or regular sets of major or of minor extension.

How are these structures formed ?

Within sedimentary rock sequences the bedding planes which define at the top and at the bottom each layer of rock, are clearly inherent to sedimentation processes (stages) which, during the geological history, have produced a given sedimentary sequence.

Some decreases in volume, because of loss of water content or because of cooling, are another cause which produces the rock separation: I refer to mudstones and to mud cracks, to joint in loëss, to the cooling joints produced in crystalline rocks, to the columnar joints in basalts.

Thus the main cause of potential and of effective separation in the rocks are the stresses of tectonical origin, which are generally slowly applied to rock masses, but sometimes may also rapidly load the rock system, as, for instance, by earthquakes or by fault processes.

We distinguish tension joints which form perpendicularly to forces tending to pull the rock apart, from shear joints which are due to forces tending to slide one part of the rock past on adjacent part.

However, in most cases, it is difficult to ascertain the origin of joints, and even if this can be done, and the attitude of the stress axes, and strain axes can be established, it may be impossible to deduce with satisfaction the character of the external forces.

An ideal example of tension joints is a cooling horizontal sheet of basalt.

The basalt solidifies at about 1.000 °C and it contracts during subsequent cooling. The resulting tensional forces act primarily in the horizontal plane and are equal in all directions within this plane; when rupture eventually takes place three vertical fractures, making angles of 120 degrees, with each other radiate out from numerous centers. If the centers are evenly distributed, the fractures bound vertical hexagonal columns.

On the contrary tension fractures due to a couple are represented by some of the crevasses in glaciers and by feather joints.

Vertical crevasses are diagonal to the contact between glacier and rock walls.

The friction with the walls sets up couples. The intermediate axis of the strain elipsoid is pependicular to the surface of the ice and the greatest and the least axes are respectively A-A' — C-C'. The **tension cracks** develop at right angles to the greatest strain axis A-A'.

Feather joints are moreover tension fractures related to faulting.

Tension joints which form perpendicularly to A-A' may be confined to one side of the fault if the rock material at that side has a lower tensile strength than the rock material on the other side.

Extension joints can be found perpendicular to the axes of folds. They are common in orogenic belts. They result from slight elongation parallel to the axes of the folds. They would be analogous to the ruptures that form parallel to the sides of specimens under compression.

Joints parallel to the axial planes of folds may be **release joints** similar to those that form at right angles to the axis of compression when the load is released. Other joints with this attitude may be due to tension at the convex side of a bent stratum.

Shear joints are difficult to recognize. If a joint is slicken-sided the opposite walls have obviously slipped past one another. But this is not proof that the fracture originated under shearing stresses. The stress may be tensional and the sliding of walls past each other could be a later phenomenon. Two sets of joints that intersect at a high angle to form a conjugate system are often considered shear fractures especially if they are symmetrically disposed about the strain axes. The joints could be interpreted as shear fractures, that developed due to compressive forces with easiest relief in a direction normal to them (compr. forces).

In other instances shear joints can result from a couple. Although many explanations for origin of joints are plausible and account for all the factors, other interpretations cannot always be eliminated.

On the mechanic of joint genesis many interesting investigations have been performed up to date. A recent comparative study performed on sedimentary rocks by Bock, based on geological and petrofabric analysis, on finite elements calculations to investigate the mechanical relationship important for tensile joint genesis, and on qualitative experiments using equivalent specimens, lead to the conclusion that for the fundamental joint system of the sedimentary rocks (two vertical joint sets) a satisfying geological and mechanical interpretation does not exist.

This problem is very interesting from a geomechanical point of view, and this subject should have to be developed because the study of deformative and rupture natural processes provides the most important conclusions on the rock mass deformation and rupture theory.

In addition to the discussed considerations on the origin of geological discontinuities, in studying the mechanical behaviour of a rock mass, the quantitative and qualitative description of rock structures is fundamental to perform a complete classification of the rock mass.

First of all we must emphasize that, when we speak about geological discontinuity, we refer to a more or less plane or to a more or less extended surface, which corresponds to an actual discontinuity in the rock substance, i.e. a break in the spatial continuity of the rock mass.

In correspondence to these surfaces we assume that the mechanical strength of the rock material has been overcome. In some instances, as we have seen, shear strength has been overcome, in some other tensile strength.

In many cases within the structure of crystal grid and in the overall order of material fabrics, even if potential discontinuity conditions are existing, real conditions of material separation have not been produced.

The surfaces of geological separation, which are commonly called joints, could have irregular diffusion and spreading attitude, or could have a well ordered attitude, in which well defined symmetry conditions can be recognized.

Compared with a more or less complex or with a more or less continuous net of geological discontinuities, a rock volume could be characterized by different conditions of separation.

Referring to a specific scale of the geomechanical problem, and according to the peculiar development of the threedimensional set of geological discontinuitites, we may distinguish different structural faces.

According to L. Müller we can define a **"multiple body system"** when the network of discontinuities has completely developed and **rock bodies** (unit rock volumes) come out to be wholly defined by a complete intersection of discontinuity surfaces.

On the contrary we can define a **"partial body"** system rock mass appears only partially subdivided. in this case unit geological bodies will be only partly defined (see sketch "b" of the projected slide).

If rock mass appears without discontinuities, we shall speak of a unit body system.

The mechanic of the continuum and the rheology will be applied to the case "a". Geomechanical, i.e. the Mechanics of the discontinuum and the science of geological structures (Gefügekunde) will be suitable to the cases "b" and "c"; Soil Mechanics and sometimes the science of geological structures are finally suitable to the case "d".

What has been said on the most typical rock mass faces, on one hand emphasizes the "relativity" of each description, definition or classification compared with the effective scale of the considered rock mass; on the other hand it shows immediately which **"structural parameters"** are to be collected and introduced into the complex system corresponding to the particular material, whose mechanical behaviour is going to be investigated.

Let us draw up a list of these structural determinations and parameters:

1. Relative classification of geological discontinuities, and determination of threedimensional orientation, i.e. determination of the attitude of geological surfaces belonging to each joint set we can recognize in the interior of the considered rock mass volume.

2. Determination, for each set of discontinuities, of the peculiar **degree of jointing.**

3. Determination, for each set of geological discontinuities, of the **joint spacing.**

4. Determination of the **degree of separation** of the rock mass.

5. Determination of the resulting form and of the dimensions (i.e. volume) of the **unit rock block,** taking into account the mutual interlock of each joint set crossing the considered rock volume.

6. Description of morphological and geometrical properties of joint surfaces, referring to each joint set considered. As we shall see later on, these properties, for instance the overall continuity, of the outcrop of a geological structure, or the continuity and the evenness in detail of the joint planes.

7. The determination of the opening of every joint belonging to each set, and the

survey of eventual fillings which could be present in the joints. Sometimes these fillings are of clayey or of silty nature, and because of their thickness must be studied in detail according to Soil Mechanics methodology.

8. The geological discontinuities must be finally studied as actual or potential water pathes or as water containers. Water represents the liquid phase in the mechanics of the rock mass, and therefore the presence of water in the joints or its percolation through joints is an important parameter which must be collected and introduced into the physical model of the rock mass, as we shall see later on (Prof. Louis' lecture).

The quoted determinations and parameters decisively allow to define **quantitatively** the presence of geological discontinuities in the geomechanical problem, both for an immediate application of structural parameters compared with "**classification**" of rock faces, which is directly connected with a prevision of the mechanical behaviour of the rock mass, and, in general, for a real and corrected foundation of a Rock Mechanics research.

The heterogeneity of attitude, the large variation of density and of geometrical characteristics which is peculiar of a net of geological discontinuities, suggests that geological structures have to be studied according to a statistical methodology.

This statistical methodology allows the sampling of averaged parameters, which can be correctly introduced into the analysis of the mechanical behaviour of the rock mass.

The statistical appraisal of vectors and of scalar quantities allows to gain the "**rules which define and govern**" the rock mass system, without however neglecting eventual deviations from statistical rules. The average statistical values, when referred to specific boundaries, represent the needed rules. Instead of having a high number of single data, we shall dispose of a limited number of collectives statistically obtained. For example, within the limits of a graphical statistical determination, it will be possible to group single geological discontinuities or associations, or those rough ones, by determining the maximum concentrations and their characteristic variations.

The validitiy of statistical methodology is strengthened by all those cases where, thanks to it, it is possible to go back to a certain structural system even there, where such a system looks quite absent.

It should be emphasized that measuring and sampling may be conducted on exposed rock faces in various forms. If the surface is large, the number of joints exposed may also be large and some form of selection may have to be applied.

Anyhow the sampling can be performed as "area" or "line" sampling (Robertson-Piteau).

The accuracy of these different methods is extensively discussed by these Authors.

From sampling depends in particular the reliability that can be placed on the joint set properties determined. In most cases the area sampled is limited by physical limitations (for instance from sampling at exposed slopes to sampling at bore holes).

The joints measured (the sample) which may be only a portion of the joint exposed (the sample population) are considered to be representative of the joints within the entire rock mass (the target population).

We understand now why in the ambit of structural Rock Mechanics we must always refer to the statistical survey of geological details.

Let us consider different geological-structural parameters which directly and decisively intervene in the quantitive classification of the rock mass.

First of all I must mention the "relative classification" of geological discontinuities. Practically this classification is based on the dimensions of geological discontinuities, as referred to a scale directly connected with the actual dimensions of the considered rock mass volume, namely to a scale directly connected with the actual dimensions of the geomechanical problem. (see Fig. 4).

Once, the classification of linear dimensions of structures has been performed, we have at our disposal a means which allows a preliminary arrangement of the structural system, and which permits the beginning of the survey of further structural parameters, proceeding structure by structure and set by set.

The attitude of a geological discontinuity in the space can be completely defined by the geological and by the "dip" of the geological plane, which defines a geological discontinuity.

In the praxis, by means of a geological compass, the threedimensional orientation of the line of dip referring to the azimuth "α" (which is the angle between North and the projection of this line), and the altazimuth "β" (which is the angle between the horizontal line and the line of dip.

Indirect determinations of the threedimensional attitude of geological discontinuities could be moreover performed when rock outcrops are not accessible to operator. (see Fig. 5).

Geological direct or indirect joint survey provides a large number of attitude values, which must be now elaborated and interpreted in order to find the rules of the spatial distribution of joint into the discontinuities net.

The only way in which the spatial distribution of joints can be described with uniformity of all features, is on a referenced sphere. Since it is desirable to present this information on a planar surface, some form of projection is used with the loss of some of the basic data relationship.

The use of Wulf and Schmidt projections are commonly made for the presentation of joint survey data and has been adapted for computer printout.

The use of these **stereoplots** is a powerful tool in the analysis, since they preserve some of the properties of the geometric relationship between the joint planes.

Sometimes rectangular plots are used. They are particularly suitable for computer constructions. (see Fig. 6).

The use of a plot which can be produced by a computer, and which displays the data in a form in which it can be readily assessed by visual inspection, considerably increases the volume of data that can be processed. (see Fig. 7).

It is apparent that the statistical determination of the maxima of concentration of joints, in structured diagrams connected with Schmidt's resolution of statistical investigation, gives us the possibility of establishing at the same time the degree of dispersion and of scattering of the found values compared with values at maximum concentration. Such determination is particularly important and meaningful in the study of plane structures properties and, after all, in the study of mechanical behaviour of the rock mass: the more limited and circumscript the scattering is in the ambit of a maximum of concentration, the more regular and well defined the structural order will be. On the contrary, the more irregular and spreading are the attitudes of joints, the more irregular and changeable the structural characteristics of the rock mass. (see Fig. 8).

The determination of the average statistical values of attitude referred to each set of joints, permits to know the spatial geometrical relations which intervene among the several elements which form the structural system, and to get at least some information on the natural stresses which have originally deformed the rock mass, and, finally, to know the relations intervening between the direction of the over-mentioned stress (or of stress artificially applied) and the attitude properties of the structural net.

As we did with the attitude of geological structures, statistical methodology is applied to the study of the other scalar quantities being included in the definition of plane geological structure.

Degree of jointing   The determination of the degree of jointing is very important in Geomechanics, since it allows to derive considerations on unit rock block dimensions, on rock mass hydraulics, on rock mass deformability, a.s.o.

Symbol "κ" refers to number of intersections of a particular set of geological discontinuities along a sampling line, whose length can be equal to 1 meter:

$$n/\ell = \kappa$$
(1)
$$\ell = 1m$$

where n is the number of jointing intersections, κ is the degree of jointing index, according to Stini.

The degree of jointing can be determined for each set of joints, when the sampling line is orientated normally to the planes of the discontinuities of this set. (see Fig. 9).

Spacing of joints   The spacing of joint denotes the average spacing among joints for sampling length. It is expressed as:

(2)
$$d = \frac{1}{\kappa}$$

In the survey praxis the minimum and maximum value of joint spacing are to be practically determined.

A statistical determination of the value of average spacing, performed on a large number of discontinuous of a particular set of joint, allows us to calculate indirectly the degree of jointing "κ" when this determination can not be made directly along a certain sampling line. (see Fig. 10, 11, 12).

Degree of separation of the rock mass   Let us now consider how the "**degree of separation**" of the rock mass can be determined.

The quantification of the degree of separation allows us to emphasize clearly the morphological anisotropy of the rock mass, which is bounded, for instance, with dimensions of rock volumina, rupture pattern, rock mass compressive strength, internal mobility, shear strength, tensile strength, the last ones in connection with actual or potential rupture surfaces.

In the praxis it is necessary to determine first twodimensionally, and then threedimensionally, the total amount of the rock mass area, where actual conditions of separation exist, comparing this value to the whole rock mass area, where effective conditions of continuity in the material are present (the so called "gaps").

We have seen how the conditions of separation are produced by joints (i.e. cracks or fractures, or diaclasses) which spread out in the rock mass grouped into sets having different average attitude.

To simplify, let us try to determine the "degree of separation" due only to one set of joints. The reasoning could be done first at two dimensions, referring to an hypothetical planar section whose surface lies on the same plane of the plane of any discontinuity of this set. This hypothetical planar section crosses completely the considered rock volume: its area is quoted as "A". The ratio between the area of the hypothetical planar section and the area of the discontinuity plane will be a ratio between a continuum and a discontinuum. (see Fig. 9).

$$x_e = \frac{K}{A} \tag{3}$$

In this specific case, the mentioned ratio "x_e" is 0.5, that means 50% of the hypothetical planar section crossing the rock mass, is separated. In this case the "gap" corresponding to the zones where the rock mass is intimately continuous, since it is actually not separated, corresponds likewise to 50%.

Let us now consider the problem in threedimensions, taking into accoung the whole rock volume represented on the projected slide.

If we compute the total area corresponding to geological discontinuities belonging to the considered set, within the given rock volume, then we may determine the total amount of the rock sections where actual conditions of separation exist, and we may compare this value with the total amount of the rock mass volume. (see Fig. 13).
We shall have:

$$x_e \cdot K = x_R \tag{4}$$

where "k", as we have seen, is the "degree of jointing". "x_R" definitely represents the actual separation produced by the joints of one particular set within a given rock volume.

In reality, on the hypothetical section we have considered (quoted with

"P$_2$" in Fig. 13 more than one discontinuity can exist, and therefore the area of each singular discontinuity that lies on the plane of the considered hypothetical planar section, must be summed up to determine the requested ratio:

$$(5) \qquad x_e = \frac{\Sigma K}{A}$$

when the total separated area is equal to the area of the hypothetical reference section, case K$_1$ + P$_1$ then the mentioned ratio becomes equal to the unit

$$(6) \qquad x_e = 1$$

In this case the rock mass along the reference section is completely separated.

Being $x_e = 0$ the rock mass, on the contrary, along the quoted section, is completely intact, that means continuous.

The two- and the threedimensional determination of "degree of separation" must be successively extended to the plane structures of each set apparent within the rock mass, so that, finally, the obtained "index" provides a quantitative parameter suitable to define precisely the separation ratio of the rock mass and the variation of the geomechanical behaviour caused by geological structures compared with different directions in the space.

A high degree of separation referred to a given reference section means, for instance, a low shear strength on the plane of this section or a low tensile strength in a normal direction to its plane due to the fact that the "gap" of virgin rock material which produce cohesion strengths, are of reduced dimensions of the separated area.

Friction strengths and, only partially, shear strengths, are on the contrary, mobilized along the planes of the discontinuities.

The concept of "boundary residual strength" (Restverband) which is peculiar in the definition of a rock mass, according to the prepositions of the Salzburg's School, is intimately connected with the presence of the mentioned gaps built up of intact rock slices.

Finally a high degree of separation means both an elevated overall mobility (i.e. deformability) of the rock mass, and a low residual strength.

The survey of geological details and parameters, to be used to determine the degree of separation of a given rock mass, involves quite delicate measurements and estimates on which lies the reliability of the displayed data.

The survey must be made, as usual on the basis of a statistical methodology with reference to geometrical parameters of each set of geological discontinuities, and should be performed by trained personnel.

Unit rock block   The unit rock volume refers to the smallest homogeneous **rock unit** produced by mutual intersection of systems of geological separations crossing a rock mass.

Therefore a rock mass could be considered to be composed by a multitude of unit rock volumes. Their representative shape and dimensions may be determined by means of statistical evaluation of attitude and spacing of joint sets. The average volume of unit rock block is:

$$V_{UB} = \left(\frac{1}{Ka}\right) \cdot \left(\frac{1}{Kb}\right) \cdot \left(\frac{1}{Kc}\right) = (da) \cdot (db) \cdot (dc) \tag{7}$$

The geometrical shape and the spatial attitude of the unit rock block can be exactly determined according to a graphical method which has been described in detail by L. Müller. (see Fig. 14).

This figure shows two typical unit rock blocks graphically constructed by Onofri, according to structural parameters belonging to stratification planes coupled with two transversal sets of major joints recognized within the rock mass.

In nature, the most recurring shape of the unit rock block are the following ones: (see Fig. 15).

When the ratio d_1/d_3 and d_2/d_3 is less than 1:5, the resulting unit rock block has a narrow and elongated form. In this case the deforming processes of the rock mass may principally be such as loosening and overturning of singular rock sectors: mutual sliding along the most elongated faces may also be possible.

When the ratio d_1/d_3 and d_2/d_3 is, on the contrary, larger than 5:1, the resulting unit rock blocks have a flat and thin form like flag-stones.

In this case the deforming processes of the rock mass may be such as mutual sliding along the flat and sometimes even face of the rock body.

Intermediate shapes (the ratio is then 1:1) produce partial and mutual rotations, a kind of rolling of unit rock bodies in which sliding and overturning are equally coupled.

The determination volume, of shape and of spatial attitude of the unit rock block is very important and meaningful in the description of the rock mass, both for practical and for theoretical purposes.

The more discontinuous (I mean separated) the rock mass is, the more significant is the concept of "**unit rock block**", because the more reduced are the actual bound conditions within the rock mass, and the rock system approximates to a collective build up of completely defined rock bodies.

Morphological properties of planes and opening of geological discontinuities
In order to derive basic considerations and conclusions on the mechanical significance of geological discontinuities and, finally, on the mechanical behaviour of the rock mass, the description of morphological peculiarities of planes of geological surfaces is of utmost importance. A qualitative-quantitative description of surface details give us information especially on potential shear and friction strength compared with particular direction of stress application or, more generally, compared with the whole behaviour of the rock mass itself.

In the praxis the overall shape of joint outcrops must be recognized and described quantitatively on a large scale (see Fig. 16).

This schematization furnishes some additional details to emphasize the therms of the quoted problem.

On the scale considered in the sketch, since the undulations of the joint surface are of such magnitude, that they are unlikely to be sheared off during failure of the rock mass along the joint, they modify the direction of initial vector of movement. On account of this, the connected geostatical problem results to be also substantially modified if compared with probable unsafe conditions produced by the theoretical sliding surface.

Waveness is measured in the field survey in terms of the amplitude and base length of the waves.

If morphological determinations are applied to smaller scale, that is at the dimensions of one square meter of few square decimeters, then the peculiar characters of plane details are described.

At this scale the determinations concern the smoothness or the roughness of surfaces and the detailed description of indentation.

The dimensions of undulations and indentations mut be, as usual, compared with the overall scale of the considered geomechanical problems. By certain values of this ratio, undulations or indentations can assume an utmost importance in modifying (increasing) the strength parameters of the rock mass, mobilizing true shear parameters of intact gaps. On the other hand, sometimes, in spite of the presence of indented geological structures, the overall strength of the rock mass results to be not increased because of the influence of weaker lithological members or of smooth joint surfaces having the same attitude, which are present within the rock mass, and which govern its mechanical behaviour.

In such cases the rupture process does not develop along sections coinciding with ondulated or indented structures (pathes of maximum shear strength), but it follows the paths of least mechanical resistence, which can even correspond to new-formed rupture surfaces.

A complete discussion on these arguments will be presented in Renger's paper.

These considerations give an account of the importance of performing complete determinations of structural morphological details, in order to gain correct information on the levels or on the directions of maximum or of least rock mass strength.

A further important determination concerns the "opening" of geological discontinuities.

As we know, the opposite lips of a joint or of a bedding plane may sometimes not be kept in intimate touch, but detached somehow.

In these cases the contacts realize sometimes only at some few points within the plane structure mostly corresponding to undulations or indentations prominences. (see Fig. 17).

When undulations are present at a structural plane, mutual little dislocations of the opposite rock blocks (i.e. fault processes) may emphasize the opening of the geological structure.

Anyhow, when a geological discontinuity is open, the total area corresponding to actual contact points, results to be always reduced. This circumstance may be connected with a reduction of potential shear strength along the given plane.

Sometimes the opposite lips of a joint or of a bedding plane are not in contact at all. In these cases they are separated by some filling material, which can, for instance, be crystallized solutions of quartz (SiO_2) or of clacium carbonate ($CaCO_3$) or of other minerals or clayey or silty materials.

When filling materials mobilize low mechanical strength (for instance marls, marly clays, clays, mylonitic clayey materials, talcous materials, silts, a.s.o.) and when their average thickness exceeds the dimensions of plane asperities, then the mechanical strength of filling materials governs the overall mechanical strength of the rock system compared with a given filled geological discontinuity.

The opening of a geological discontinuity is expressed in millimeters. The overall value of opening determined for each given set of geological structures and, finally, for the whole rock mass is, once again, a statistical parameter.

This quantitative parameter is very meaningful because it gives an account both of considerations referred to the degree of loosening of the rock mass, and of considerations connected with water, standing or percolating within the rock mass (Louis' lecture).

The more elevated the total value of opening, the more "losened" the rock mass will be. The concept of degree of loosening of the rock mass is directly connected with considerations on the degree of compressibility, in the whole or, at least, towards some specific directions, when, for instance, accentuated opening refers to only one particular set of geological discontinuities.

Moreover, the higher the resulting value of overall opening, the greater the water volumes within the rock mass which can produce idrostatic or hydrodynamic pressures modifying substantially the mechanical behaviour of a given rock system. In some instances the average opening of one set of joints, compared with perpendicular degree of freedom of the rock system, may actually produce orientated water pressures which could be of paramount importance in geostatical problems.

Homogeneity zones   It is possible to recognize that in the field of Mechanics of jointed rock, a rock mass may be divided itno zones of similar structural properties, for instance similar jointing, which can be finally visualized as structural regions.

For example, since jointing to a large extent controls the mechanical behaviour of rock masses, these structural regions are zones of similar strength, deformability, hydraulic, i.e. they are, in a statistical sense, homogeneous with regard to strength, deformability, and hydraulic, However, this requires that jointing in a structural region must not only be similar in such factors, as roughness, waveness, thickness, type of fillings, potential or effective water percolation, a.s.o.

A detailed statistical investigation of all structural parameters furnishes sets of quantitative parameters whose processing permits to fix the boundaries of more or less well defined homogeneity zones. Each defined zone has peculiar geomechanical characters. For each zone we can derive classifiable details.

Only within the boundaries of the given homogeneity zone we are allowed to extrapolate the determined structural and mechanical parameters.

On that account the definition of limits and characters of the homogeneity zones is the most important stage of the rock mass description. This stage usually goes on in parallel lines with the working out of data displayed by field structural surveys.

GEOMECHANICAL CLASSIFICATION OF THE ROCK MASS

We have tried to present some general aspects of the incidence of Geology in Rock Mechanics at the different dimensional scales of the problem, that is from crystalline associations to the actual rock mass, as considered in most practical purposes.

Geological parameters, both lithological and structural as derived from field surveys and conveniently worked out, are suitable to be combined to produce classifications of the rock mass at each scale of the problem, having peculiar geomechanical meaning.

This subject, which is also connected to technical properties of materials and rock masses, will in particular be handled by Prof. L. Müller.

Therefore we confine ourselves to some anticipations dealing with some structural-mechanical concepts, on whose base a geomechanical classification can be established. (see Fig. 18).

This diagram shows how the mechanical resistence of the rock mass can be determined combining quantitative data of joint frequency (abscisse), and qualitative parameters concerning the degree of weathering. The values refer to singular homogeneity zones $Z_{01} - Z_{02}$. The circular diagram indicates the areal significance of the given homogeneity zone compared with the whole rock mass area. If we combine the quantitative parameters of joint frequency with the mechanical resistence of rock material (on the ordinate, expressed in Kg/cm^2), then we obtain a quantitative classification of the rock mass compared with its mechanical strength.

The strength of rock material is expressed in Kg/cm^2, the joint spacing in cm. The zone having the greatest strenth is clearly the Z_{03} one   (high material strength − scarce spacing).

The diagram in the upper half of Fig. 19 allows to determine the partial mobility of the rock mass in relation to spacing of joints and degree of separation.

In the diagram are bounded zones where the rock mass is quasi-monolitic (zone I) or completely subdivided (zone IV). The spacing of joints is expressed in centimeters.

The diagram in the lower half of Fig. 19 shows the lines of equal mass strength depending on strength of rock material and spacing of joints.

Conditions of strong rock mass are those of zone A: the rock mass is scarcely jointed and is built up of very sound rock material.

Conditions of very weak rock mass are those of zone D (very closed joints combined with soft rock material).

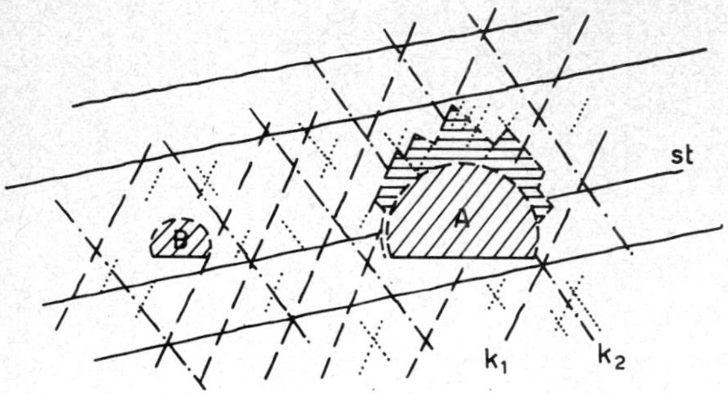

Fig. 1 Example where the structural conditions are referred to the "dimensions" of geotechnical problem. The scale of the tunnel "A" and of adit "B" will closely influence the geomechanical response of the rock mass to excavations.

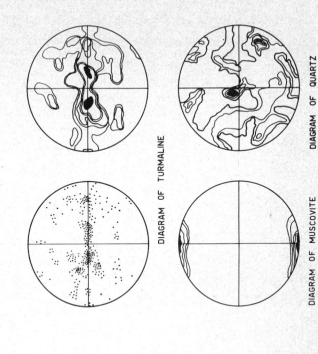

DIAGRAM OF TURMALINE

DIAGRAM OF QUARTZ

DIAGRAM OF MUSCOVITE

Fig. 2

Fig. 3 Dimensions of the rock sample commonly used for laboratory determinations on "rock material"

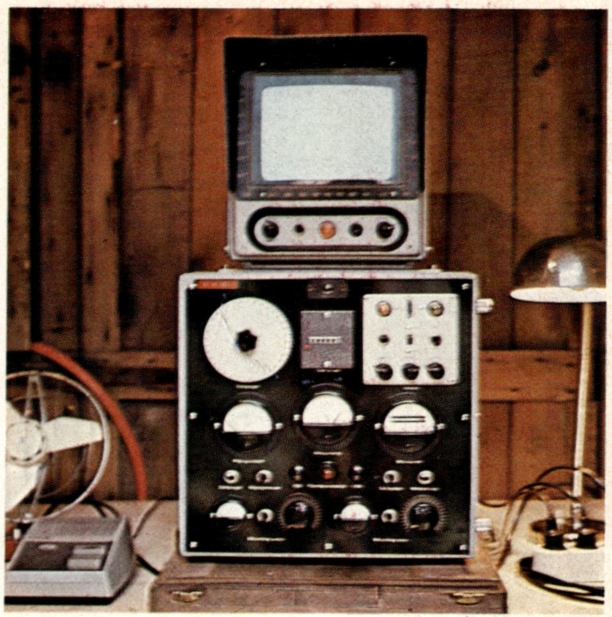

Fig. 5 The television probe developped by L. Müller, for indirect measurements of lithological and structural details in bore holes.

ASSOLUTE CLASSIFICATION OF JOINTS

DENOMINATION	SYMBOLS	EXTENSION (m)
MINOR JOINTS	k	< 1
MAJOR JOINTS	K	1 - 10
LARGE JOINTS	R	>10

RELATIVE CLASSIFICATION OF JOINTS

	DIMENSIONS (m)	DIAMETER OF INFLUENCED ZONE (m)	EXTENSION OF JOINTS (m)		
CIVIL CONSTRUCTIONS	L	D	k^{*}Rel	K^{*}Rel	R^{*}Rel
ADITS AND SMALL FOUNDATIONS	Ø=3 b=3	10 10	0-0,2	0,2-2	> 2
TUNNELS SLOPES	Ø=30 h=100	100 100	0-2	2-20	>20
CAVERNS SMALL DAMS	h=40 h=40	>100 >100	0-2,5	2,5-25	>25
LARGE DAMS LARGE SLOPES	h=100 h=300	300 300	0-6	6-60	>60

Fig. 4 Absolute (above) and relative (below) classifications of geological structures. The relative one is referred to scale of geomechanical problem.

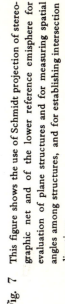

Fig. 7   This figure shows the use of Schmidt projection of stereographic net and of the lower reference emisphere for evaluation of plane structures and for measuring spatial angles among structures, and for establishing intersection directions.

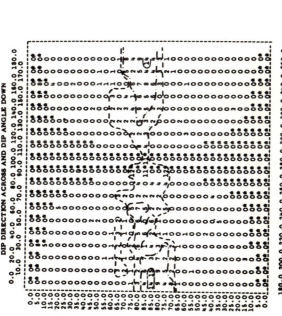

DESIGN JOINT SURVEY FOR DE BEERS MINE
OF ALL JOINTS

SECTION 13   PLOT CORRECTED FOR DIRECTIONAL BIAS

Fig. 6

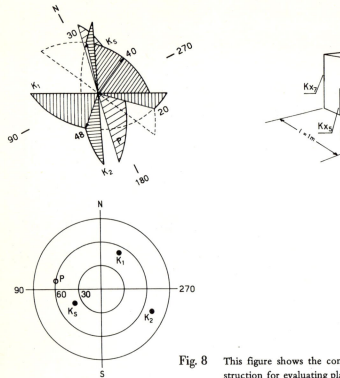

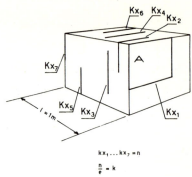

Fig. 9

Fig. 8   This figure shows the common use of great circles construction for evaluating planar relations among structures,
for instance in geostatical problems.

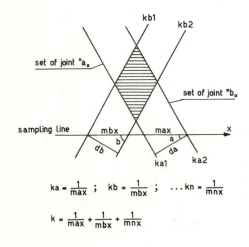

$$ka = \frac{1}{max} \; ; \quad kb = \frac{1}{mbx} \; ; \quad \ldots kn = \frac{1}{mnx}$$

$$k = \frac{1}{max} + \frac{1}{mbx} + \frac{1}{mnx}$$

Fig. 10   This figure shows as the
average total degree of
jointing "k" can be computed referring to the number
of sections of each joint
set per unit lenght.

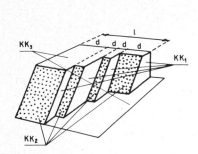

Fig. 11

Fig. 12   Regularly spaced joints at an underground outcrop.

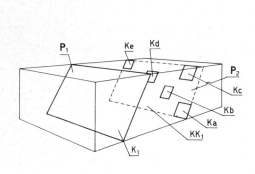

Fig. 13

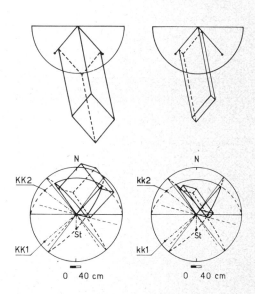

Fig. 14

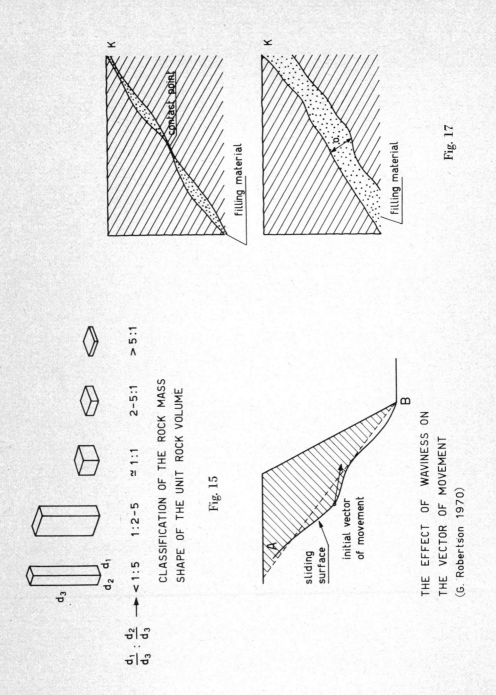

$\frac{d_1}{d_3}; \frac{d_2}{d_3}$ →   <1:5   1:2-5   ≅1:1   2-5:1   >5:1

CLASSIFICATION OF THE ROCK MASS
SHAPE OF THE UNIT ROCK VOLUME

Fig. 15

Fig. 17

THE EFFECT OF WAVINESS ON
THE VECTOR OF MOVEMENT
(G. Robertson 1970)

Fig. 16

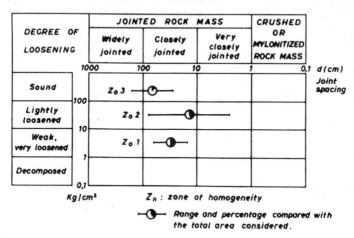

Fig. 18

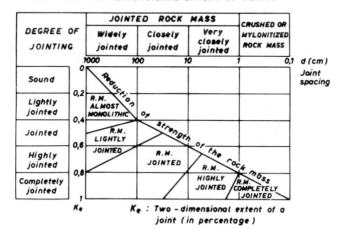

Fig. 19

A REVIEW OF FRACTURE CRITERIA OF BRITTLE ROCK

F. RUMMEL
Institut für Geophysik
Ruhr-Universität Bochum

---

* All Figures quoted in the text are at the end of the lecture.

Introduction

Much of the research on fracture of rocks or rock-like materials conducted over the past two decades may be considered as "academic studies" of the general phenomenon of fracture. Yet, the understanding of this phenomenon is fundamental if a material is used in any engineering design, whether the aim is to prevent failure of the structure or to promote it.

Fracture theories existing are generally empirical and derived from experimental results of laboratory test with simple boundary conditions. Because of the basic weakness of rock intension and because in general the environmental stresses in rock mechanics are compressive most of these theories consider fracture under compressive stress conditions. The Coulomb-Navier-, the Mohr-, the Griffith and the McClintock and Walsh criteria are typical examples and will be considered in the following. In addition the tendency during the past was in making accurate experiments under conditions of homogeneous stresses. To obtain information about the fracture behaviour with unequal principal stresses systems have to be used which involve inhomogeneous stresses. This case is of particular interest, since in practical rock mechanics we may expect conditions of highly inhomogeneous stresses. However, a consideration of such situations involve additional assumptions like the applicability of the theory of elasticity for calculating the stress field, which may be open to question. A distinction has to be made between fracture initiation and fracture propagation, since a detailed observation of the total fracture process in rock was possible by means of "stiff" and "servo-controlled" loading systems. In general the criteria mentioned above only consider the initiation of fracture such as the moment of crack instability rather than crack growth.

There already are a great number of comprehensive reviews devoted to brittle fracture of rock. During the last decade such reviews have been presented by JAEGER (1966), HOEK (1968), BRACE (1967) and most recently by FAIRHURST (1971 and 1973). The present review is essentially based on the articles mentioned above and does not claim to be comprehensive in any respect.

Notation

In opposite to the convention adopted in continuum mechanics compressive stresses will be reckoned positive, where the principal stresses $\sigma_1 \geqslant \sigma_2 \geqslant \sigma_3$.

It is assumed that failure takes place when a definite relation $\sigma_1 = f(\sigma_2, \sigma_3, \text{material})$ is satisfied. This relation will be called criterion of fracture. For brittle fracture failure may be assumed to take place at the maximum value of σ_1, which is called the strength of the material (Recently often the term "maximum load-bearing capability" is used to indicate that the fracture process may be described as a continuous progressive structural breakdown of the material). This conception was generalized by JAEGER into the statement, "that if any particular component of stress is increased under specified conditions until failure occurs, the magnitude of that stress of failure is known as the strength of the material under those conditions." With this conception we may speak of "uni-axial tensile or compressive strength, triaxial compressive strength" at a certain constant confining pressure etc.

Most discussions on fracture start from ideal models such as homogeneous, isotropic, elastic materials, for which two possible fundamental fracture mechanisms, **shear** and **cleavage**, are discussed. Both of these mechanisms are based on the possible motions of atoms or molecules within the crystal lattice. Shear may be considered as the dominant deformation (or fracture) mechanism for the highly ductile behaviour of rock at high confining stresses and higher temperatures. It is generally considered from the theory of dislocation point of view. Cleavage fracture may be characterized by an increase of the interatomic spacing to such a distance that the atomic bonds acting across the cleavage plane are negligible. Cleavage is generally observed for brittle rock or better brittle conditions such as low confining pressures and low temperatures.

The macroscopic fracture phenomenon observable in rock such as frictional sliding or crushing of material can be reduced to an interaction of these two basic mechanisms.

The description of the makroscopic or microscopic fractured surfaces is of great importance in relation to the criteria applied. Various **types of fractures** are illustrated in Fig. 1, which may f.e. demonstrate the behaviour of Solnhofen limestone at various stress-strain conditions. The two basic fracture types, shear and cleavage fracture can be clearly seen.

Coulomb-(Navier) Criterion

The simplest and certainly most important criterion is the "internal friction" criterion associated with the names of COULOMB (1776) and NAVIER (1883). Coulomb studied the shear fracture in prisms of quasi-isotropic materials under uniaxial compression σ_1. From his experiments on rocks he suggested that the shear stress causing failure along a plane inclined to the principal applied stress is obtained by

(1) $$|\tau| = \tau_o + \mu\sigma$$

where τ and σ are the shear and the normal stress in the failure plane, τ_o is often referred to as the cohesive shear strength and μ is the coefficient of internal friction.

The normal stress σ and the shear stress τ can be given in terms of the applied principal stresses σ_1, and σ_3

(2) $$\sigma = 1/2(\sigma_1 + \sigma_3) + 1/2(\sigma_1 - \sigma_3)\cos2\theta$$

(3) $$\tau = -1/2(\sigma_1 - \sigma_3)\sin2\theta$$

where θ is the angle between the potential shear plane and the minor principal stress σ_3 (the angle between the normal to the plane and the maximum principal stress). The quantity $(\tau - \mu\sigma)$ from eq. 1. is then given by

$$|\tau| - \mu\sigma = 1/2(\sigma_1 - \sigma_3)[\sin 2\theta - \mu\cos 2\theta] - 1/2\mu(\sigma_1 + \sigma_3)$$

Fracture will occur if $|\tau| - \mu\sigma$ is a maximum and is equal to τ_o. By differentiation with respect to θ we obtain then

(4) $$\tan 2\theta = -1/\mu$$

so that 2θ lies between $90°$ and $180°$ — depending on μ and the normal stress. Writing the criterion in terms of the principal stresses σ_1 and σ_3

(5) $$2\tau_o = \sigma_1\left[\left(\mu^2 + 1\right)^{\frac{1}{2}} - \mu\right] - \sigma_3\left[\left(\mu^2 + 1\right)^{\frac{1}{2}} + \mu\right]$$

Eq. (5) is a straight line in the σ_1-σ_3 plane (Fig. 2) which intercepts the σ_1-axis at

(6) $$C_o = 2\tau_o\left[\left(\mu^2 + 1\right)^{\frac{1}{2}} + \mu\right]$$

and the σ_3- axis at

(7) $$S_o = -2\tau_o\left[\left(\mu^2 + 1\right)^{\frac{1}{2}} - \mu\right]$$

C_o is the uniaxial compressive strength.
(Note: S_o is not the uniaxial tensile strength!)

Considering the condition that $\sigma > 0$ eq. 5 represents a valid criterion for all values of

$$\sigma_1 > \tau_0 \left[(\mu + 1) + \mu \right] = 1/2 \ C_0 \tag{8}$$

If one of the principal stresses is tensile extension fractures in planes perpendicular to σ_3 may occur (f.e. in uniaxial tension at the tensile strength σ_t). This behaviour is entirely different from the shear fracture and is not described by the Coulomb criterion.

By introducing the angle of internal friction, ϕ , defined by

$$\mu = \tan\phi \tag{9}$$

into eq. (5) a very useful form of the criterion may be obtained, which allows to plot σ_1 at failure in terms of σ_3 (Fig. 3)

$$\sigma_1 = Co + q\sigma_3$$
$$q = \tan^2 \alpha \tag{10}$$
$$\alpha = \pi/4 + 1/2 \ \phi$$

Finally, the criterion may be represented in the $\tau - \sigma -$ plane as used for the Mohr's representation. In the $\tau - \sigma$-plane eq. (1) is represented by a straight line of a slope of $\mu = \tan \phi$ with respect to the σ-axis and an intercept of τ_0 , on the τ -axis (Fig. 4). The normal and the shear stresses across a potential shear plane are given by the Mohr circle. Failure will occur if the Mohr circle touches the line given by eq. (1). Then it follows that

$$2\theta = 1/2\pi + \phi \tag{11}$$

and the criterion may be stated as

$$1/2(\sigma_1 - \sigma_3) = \left[\tau_0 \cot \phi + 1/2(\sigma_1 + \sigma_3) \right] \sin \phi \tag{12}$$

The Coulomb criterion of failure does not consider the intermediate principal stress σ_2 !

In general experimental results of triaxial tests confirm the Coulomb criterion by giving a linear $\sigma_1 - \sigma_3$ relationship for many rocks within a limited stress range.

2. Mohr's Fracture Criterion

Mohr (1900) generalized the Coulomb criterion considered the triaxial state of stress. His most important contribution is that he recognized that the material properties themselves are functions of the stress. He summarized that "at the limit the shear stress on the glide plane reaches a maximum depending on the normal pressure and the character of the material", which may be expressed by the functional relation

$$(2.1) \qquad\qquad\qquad \tau = f(\sigma)$$

eq. (2.1) represents a curve in the $\sigma - \tau$ plane, which may be determined experimentally as the envelope to the Mohr circles corresponding to fracture under various stress conditions. Implicit in this criterion is that the intermediate principal stress σ_2 has no influence on the fracture, an assumption which is open to question.

If the Mohr envelope is linear, the Mohr criterion is equivalent to the Coulomb criterion even when the physical background of both criteria is different.

For material with no "cohesive shear strength" the Mohr envelope passes through the origin of the $\tau - \sigma$-plane ($\tau = \sigma \tan\phi$) , which then describes the Rankine condition used in soil mechanics. The envelope will also pass through the origin if fracture or sliding occurs along a pre-existing fracture plane. In this case the slope of the $\tau - \sigma$-curve determines the coefficient of external (or sliding) friction (in opposite to the coefficient of internal friction for intact rock).

In addition another simple case may be mentioned, that of a parabolic Mohr envelope, which may be obtained by the Griffith theory.

3. Griffith Criterion
(following Fairhurst 1971)

The tensile strength of an ideal crystal may be calculated by considering the interatomic bonds existing within the crystal lattice. The equilibrium position is determined by the net force (Fig. 5a) which results by superposition of attractive and repulsive forces to be zero. Application of external tension will result in an increase of the atomic lattice spacing and will lead to an increase of the interatomic "stress" (Note that the mathematical concept of stress at a point is not a physical reality!).

By approximating the interatomic force vs interatomic spacing curve (or interatomic stress vs interatomic spacing curve) (Fig. 5b) by a sine curve

$$\sigma = \sigma_{id} \sin \frac{2\pi d}{\lambda} \qquad (3.1)$$

(σ_{id} is the ideal interatomic tensile strength, d is the atomic spacing, and λ is a parameter describing the force interaction within the solid) a rough estimation of σ_{id} is possible. The approximation leads to

$$\sigma_{id} \approx \frac{E^*}{2\pi} \approx \frac{E^*}{10} \qquad (3.2)$$

where E^* is the intrinsic Youngs modulus of the solid. A comparison of the ideal values for the tensile strength obtained by 3.2 with experimental values for rocks, leads to a discrepancy of several order of magnitudes. For example for Solnhofen limestone:

$$E \approx 5 \cdot 10^5 \ \frac{kp}{cm^2} \ \left| \ \begin{array}{l} \sigma_{id} \approx 5 \cdot 10^5 \ \frac{kp}{cm^2} \\[2mm] \sigma_{comp} \approx 2 \cdot 10^3 \ \frac{kp}{cm^2} \\[2mm] \sigma_{tension} \approx 3 \cdot 10^3 \ \frac{kp}{cm^2} \end{array} \right.$$

The conclusion by Griffith (1920) was that the crystal lattice cannot be considered as an ideal model to explain the macroscopic strength behaviour of solids and in particular, of rocks. Stress concentrations generated at the tips of so-called "Griffith cracks" have to be considered. Further Griffith recognized the need to consider the energy necessary to form a fracture surface.

a) The stress concentration and thus the condition for the external stress to cause fracture may be considered by studying the stress situation in a thin, elastic, isotropic, infinite plate of unit thickness containing a single elliptical crack of length 2 c and loaded in plane stress by a constant tensile stress (Fig. 6).

The stress σ_t occuring at the crack tip is given by

$$\sigma_t = \sigma \left(1 + \frac{2c}{b} \right) \qquad (3.3)$$

where 2 b is the minor axis of the elliptical crack. For sharp cracks the induced tensile stress σ_t at the crack tip could approach the cohesive strength of the material as defined by 3.2.

b) In formulating the energy condition for fracture Griffith used the "theorem of minimum potential energy": The stable equilibrium of a system is given when the potential energy of the system is a minimum. Relating to fracture he had to add the postulation.

"the equilibrium position, if equilibrium is possible, must be one in which rupture of the solid has occured, if the system can pass from the unbroken to the broken condition by a process involving a continuous decrease in potential energy".

Since the process of passing "from the broken to the unbroken condition" occurs by progressive lengthening of the crack across the infinite plate, the mathematical formulation of the energy criterion involves a consideration of the energy change occuring during crack growth. The energy terms to be considered are the change in potential energy of the applied forces, ΔW, the change in strain energy due to the existence of a crack, U and the change in surface energy, ΔS. Thus, the total change in potential energy of the system, ΔP, is

$$\Delta P = \Delta W + \Delta U + \Delta S$$

where

(3.4)
$$\Delta W = - 2\Delta U \qquad \text{(Love 1927)}$$
$$\Delta U = \frac{\pi c^2 \sigma^2}{E}$$
$$\Delta S = 4\gamma c \qquad (\gamma \text{ is the specific surface energy})$$

Eq. (3.4) is shown in Fig. 7 for variable crack length and for two different stress levels, σ_a, and $\sigma_b (\sigma_b > \sigma_a)$.

Considering f.e. the stress level σ_a cracks with halflength $0 < c < c_a$ would be stable, since crack growth requires an increase in the potential energy. Cracks with $C > C_a$ are unstable due to the continuous decrease in potential energy of the system. Thus, the Griffith criterion for tensile fracture may be stated as

(3.5)
$$\frac{\delta \Delta P}{\delta c} = 0$$

or introducing eq. 3.5 into eq. 3.4. we obtain

$$\sigma = \left(\frac{2E\gamma}{\pi c} \right)^{\frac{1}{2}}$$

(3.6)

where σ_g is the tensile strength according to the Griffith criterion.

In eq. 3.6 E defined the Young's modulus of the plate material without a crack. By introducing a crack of half-length C we may define an average elastic modulus E* of the plate

$$E^* = \frac{AE}{A + 2\pi c^2}$$

(3.7)

(A being the cross-sectional area of the plate). Then, for a linear elastic material the stress-strain relation at the begin of crack extension as defined by eq. 3.6 is given by

$$\sigma_g = E^* \epsilon_g$$

(3.8)

(ϵ_g is the average strain parallel to the applied tension σ). Combining eqs. 3.6 - 3.8 we obtain

$$\epsilon = \frac{\sigma_g}{E} + \frac{8E\gamma^2}{A\pi \sigma_g^3}$$

which represents the criterion for crack extension in the $\epsilon - \sigma$ -plane (Fig. 8). The curve is called the Griffith locus (Berry 1960).

For large values of σ_g the curve is asymptotically approaching the straight line $\sigma = E \cdot \epsilon$, characterizing the stress-strain behaviour of the uncracked plate. With increasing crack length $\sigma_g \to 0$ as $\epsilon_g \to \infty$.

It should be mentioned that the Griffith criterion as stated by eq. 3.6 defines the begin of crack extension for tensile stresses and it does not give any statement on fracture propagation or the behaviour of a crack in compression. A modification to take this into consideration has been suggested by McClintock and Walsh (1962), known as the Modified Griffith Criterion. The important suggestion was to take the frictional forces into account which will arise when the cracks are closed by compressive stresses. It is interesting to note that the Modified Griffith Criterion is equivalent to the Coulomb Criterion for sufficiently high compressive stresses to close all cracks existing in the material.

In the application of the Griffith criterion to rocks eq. 3.8 requires known values for γ and the initial crack length. Many attempts have been made to measure the tensile strength of rocks to determine these parameters. However, only a few reliable measurements for γ are available. Usually a statistical distribution of cracks within the rock is assumed with the largest crack controlling the strength. It must be mentioned, however, that a large variability of tensile strength is observed for different test geometries and experimental boundary conditions, so that it appears that much further studies are required to fully understand this single problem.

APPENDIX

Influence of System Stiffness to Fracture Initiation

It has often been believed that the system stiffness (including the boundary) will influence fracture or better fracture initiation. It can be demostrated that this is not so, but rather that fracture initiation is independent on the load-displacement characteristics of the load application system.

Let us consider a thin plate containing a crack, being stressed in an infinitely rigid plane.

Let the stiffnesses of the two major contributions be symbolized by two springs with stiffnesses k_s and $k(c)$. F is the applied force.

Strain energy of the system U :

$$U = [k(c) + k_s]\frac{F^2}{2}$$

If dU is the change in strain energy due to the change in force applied or due to crack-lengthening; then $dU = F^2/2 \; d[k(c)] + [k(c) + k_s]FdF$

a) "soft" system analogy to a dead load situation

condition: $F = const > dF = 0$

Thus:

$$dU = \oplus \frac{F^2}{2} d[k(c)]$$

the increase in strain energy is only dependent on the crack. Since the rigid boundary moves, the applied forces perform work.

$$dW = -Fdx$$
$$x = [k(c) + k_s]F$$
$$dx = Fd[k(c)]$$

$$dW = -F^2 d[k(c)]$$

$\left(\begin{array}{l} \text{W potential energy at} \\ \text{the applied forces} \end{array}\right)$

b) infinitely "stiff" system.

   condition

$$dx = [k_s + k(c)]dF + F d[k(c)] = o$$

$$[k_s + k(c)]dF = - F d[k(c)]$$

$$dU = \frac{F^2}{2} d[k(c)] - F^2 d[k(c)]$$

$$dU = - \frac{F^2}{2} d[k(c)]$$

$$dW = o$$

c) total energy change ΔP

$$\Delta P = \Delta W + \Delta U + \Delta S$$

case a) "soft":

$$\Delta P = - F^2 d[k(c)] + \frac{F^2}{2} d[k(c)] + 4\gamma c$$

case b) "stiff"

$$\Delta P = o - \frac{F^2}{2} d[k(c)] + 4\gamma c$$

In both cases the change in total potential energy is the same. Thus, system stiffness has no influence on crack initiation. (It has on crack propagation) ! !

REFERENCES

[1]   J.P. Berry (1960): Some kinetic considerations of the Griffith criterion for fracture. J.Mech. Phys. Solids, Vol. 8, 194-206; 207-216.

[2]   W.F. Brace (1967): Review of Coulomb-Navier fracture criterion. NSF Rock Mech. Sem., Boston College, ed. by R.E. Riecker.

[3]   C. Fairhurst (1971): Fundamental considerations relating to the strength of rock. Sem. on Fracture Mechanisms, held at Ruhr University Bochum. Publ. in Veröff. d. Inst. f. Bodenmechanik und Felsmechanik, Univ. Karlsruhe, 1972.

[4]   C. Fairhurst (1973): Estimation of the mechanical properties of rock masses. Final report ARPA-USBM, Univ. of Minnesota.

[5]   D. Griggs and J. Handin (1960): Observations on fracture and a hypothesis of earthquakes, Geol. Soc. Am. Mem. 79, Rock Deformation.

[6]   A.A. Griffith (1920): The phenomena of rupture and flow in solids. Phil. Trans. Royal Soc., 221 A, 163-198.

[7]   E. Hock (1968): Brittle failure of rock. Rock Mech. publ. by Stagg and Zienkiczwicz, Ch. 4, S. 99-124.

[8]   J.C. Jaeger (1966): The brittle fracture of rocks. 8th Symp. Rock Mech., Minnesota, 1966, AIME Proc. 1967.

[9]   F.A. McClintock and J.B. Walsh (1962): Friction on Griffith cracks in rocks under pressure. Proc. 4th U.S. Congress Appl. Mech., 1015-1022.

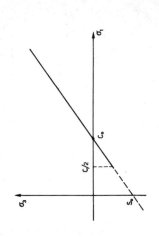

Fig. 2: Coulomb criterion in the σ_1, σ_3 plane according to eq. 5.

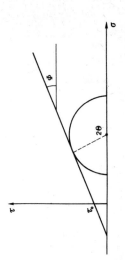

Fig. 4: Coulomb criterion in the τ, σ plane according to eq. 1.

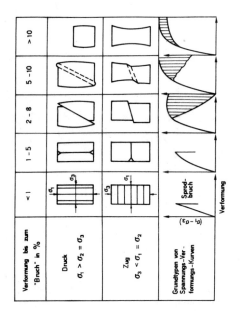

Fig. 1 : Basic types of fractures (after Griggs and Handin 1960)

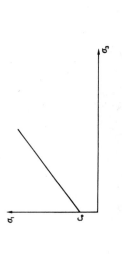

Fig. 3: Coulomb criterion in the σ_1, σ_3 plane according to eq. 10.

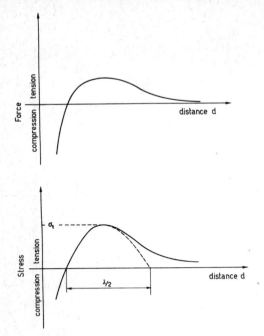

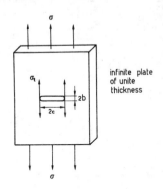

Fig. 6: Plate model containing an elliptic crack under uniform tension.

Fig. 5: Schematic representation of interatomic net forces (a) and interatomic stress (b) versus interatomic spacing.

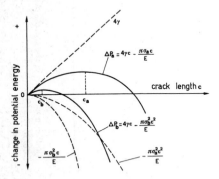

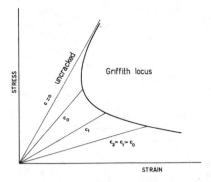

Fig. 7: Change of total potential energy, P, due to introduction of a crack of variable halflength, c, for three different constant stress levels.

$$\sigma = 0, \; \sigma = \sigma_a, \; \sigma > \sigma_b, \sigma < \sigma_b$$

Fig. 8: Griffith locus for crack instability under uniform applied tension.

BRITTLE FRACTURE OF ROCKS

F. RUMMEL
Institut für Geophysik
Ruhr-Universität Bochum

---

* All figures quoted in the text are at the end of the lecture.

Introduction

Most rocks specimens fail violently and uncontrollably at their peak strength when tested under unconfined stress conditions in conventional hydraulic loading systems. At failure their resitance to carry load rapidly drops to zero. This behaviour is commonly known as the "brittle" behaviour of rocks. Only recently it was recognized that such a behaviour is largely due to the rapid release of stored strain energy from the specimen-machine system (COOK 1965; HUGES and CHAPMAN 1966; BIENIAWSKI 1967). The understanding of this problem made possible a considerable progress in studying the fracture processes in rocks in laboratory testing. WAWERSIK (1968) showed in one of the most comprehensive fracture studies up to date by using a "stiff" loading system that fracture of rock may be considered as a continuous progressive and controllable breakdown process occurring over the entire deformation range, from initial loading to the complete disintegration. The concept of the application of such "stiff" loading systems related to rock testing is shown in Fig. 1 for a soft (K_1), a relatively stiff (K_2) and a very stiff (K_3) system (K are the stiffnesses, defined as the ratio between the drop in force, ΔF, and the corresponding platen displacement).

The influence of the stiffness of the loading system on the post fracture behaviour may also be demonstrated by considering the Griffith locus (Fig. 2). Again we have three systems with different stiffnesses 1, 2 and 4. The initial crack length of the specimen may be c_0. Crack extension (we assume tensile stresses acting at the crack tip) is initiated at A. By the very soft system 1, the stress-strain curve then follows A-1 and crack extension is associated with a considerable amount of energy release, which will rapidly accelerate the crack to high propagation velocity. Using system 2 the excess energy is less and after accelerating the crack through AB (the crack has maximum velocity at B) crack propagation may eventually come to stop at C, having now the crack length c_1 and will be stable at the stress level now existing.

A further development in the experimental study of the fracture phenomenon was achieved by making use of electrohydraulic servo controlled loading machines with fast response time, now availably (RUMMEL and FAIRHURST 1970). A review on the application of such systems is given by HUDSON et al. (1972). Fig. 3 and Fig. 4 show the principle concept.

Since then a large amount of work has been done on brittle fracture of rock using various testing situations such as uniaxial compression and tension, indirect tension in ring-Brazilian and beam tests, triaxial loading etc. A summary of

the most important experimental studies has recently been presented by
FAIRHURST (1973). Here some examples of typical laboratory fracture tests and
some field situations related to earthquake mechanisms will be discussed.

Rock Fracture in Compression

Most practical situations in rock mechanics involve compressive loads.
We may assume that brittle fracture under compressive stresses results from the
action of induced tensile stresses at stress concentration. One can easily demonstrate
this in a simple test configuration by uniaxially compress prismatic rock specimens
with a hole normal to the direction of the applied load. The test situation and the
stress distribution is shown in Fig. 5. The analytical solution of the stress field
around the opening yields induced tensile stresses in the region near the opening
with a maximum value at $r = R$ and $\theta = 0$ (r and 0 polar coordinates, R radius of
the hole). During the experiment tensile cracks will initiate at these loci of
maximum tension and propagate with further specimen deformation. Crack
propagation will be controlled by the stress field, the energy availably to create new
crack surface and the surface energy. Typical stress strain curves in both longitudinal
and lateral direction are shown in Fig. 6.

Similar results may be observed for uniaxial compression tests on
cylindrical specimens and it can be shown that disintegration of the rock occurs by
the development of numerous cracks predominantly oriented parallel to the
direction of the applied stress 1, demonstrating the presence of pre-existing Griffith
cracks in the material. Griffith cracks in the case of rocks may be formed by pore
spaces, soft or hard inclusions or grains, grain boundaries and triple points etc.,
which act to induce tensile stress region within the overall compressive stress field
generated by the applied uniform load. If the externally applied load is not uniform
the induced indirect tensile stress field may extend over a large region within the
specimen. In this situation crack propagation normal to the direction of induced
tension is enhanced, and may eventually lead to uncontrollable macroscopic
collapse (f.e. vertical splitting of uniaxially loaded cylindrical specimens). Very
typical for such a situation are indirect tension tests like beam test or the
"Brazilian" test.

While axial cracking parallel to the direction of the uniaxial load
appears to be the major fracture mechanism in most rocks, axial cracking is rapidly
depressed in the presence of a confining pressure. At increasing confining pressure
localized shear fractures become predominant and the material will behave more and

more ductile. At high confining pressure the ductile behaviour is macroscopically characterized by the development of "Lueder's lines" representing the shear planes of dislocation movement on the circumferential boundary. Typical stress strain curves for cylindrical Ruhr-Sandstone and marble specimens are shown in Fig. 7.

Dilatancy

Only recently measurements of volume changes associated with rock disintegration became of particular interest both in engineering rock mechanics as well as for earthquake prediction problems. Experiments show that the volume increases with progressive fracture. This effect is commonly called as dilation, dilatation or dilatancy. Under low confining pressures dilation $\Delta V/V$, where ΔV is the volume increase may reach a considerable amount, depending on the rock. It is assumed that it results from crack development, during increasing axial deformation. Fig. 8 shows experimental results obtained for sandstone and marble. It is interesting to mention that in marble dilation is still present at a confining pressure of 1kb. This is surprising since the transition from brittle to ductile behaviour was expected to be characterized by a marked rop in dilation due to the predominant shear. Therefore it must be concluded that a pure shear mechanism òccurs at much higher confining pressures.

Velocity Measurements

Measurement of the velocity of elastic waves may also be used in interpreting the state of deformation in rocks. Rocks commonly show an increase in velocity of a few percent with increasing stress. The onset of inelastic deformation however, is characterized by a following decrease in velocity. This effect is enhanced with increasing deformation in particular in the post-failure range. This effect has recently been used to earthquake prediction. Experimental results are given in Fig. 9.

Effect of Interstitial Pore Pressure

The effect of interstitial pore pressure on the fracture strength of porous materials is commonly described by the "law of effective stress" as stated by Terzaghi (1929) for soil mechanics. In 1945 Terzaghi suggested that the concept should be equally applicable to saturated rock. If P_p is the interstitial pore pressure and σ_1 , σ_2 , σ_3 the principal stresses the effective stresses are defined

$$\bar{\sigma}_1 = \sigma_1 - P_p$$
$$\bar{\sigma}_2 = \sigma_2 - P_p$$
$$\bar{\sigma}_3 = \sigma_3 - P_p$$

as shown in Fig. 10. The effect of the interstitial pore pressure is demonstrated in a Mohr diagram in Fig. 11. As soon the Mohr circle according to the effective stresses touches the failure envelope fracture will occur. This effect is important in earthquake triggering by pumping fluid into deep boreholes as recently shown by RAYLEIGH et al. (1972) at Rangeleg, Colorado. The negative effect when pumping fluid out of the underground was suggested as a mean to prevent slip along a fault such as the San Andreas fault. If both the effects being used this could be a possibility to trigger earthquakes along extended fault planes within small fault intervals which are separated by "stress hardened" sections (RAILEIGH and DIETERICH 1972).

These practical possibilities of application led to comprehensive laboratory studies on the influence of pore fluid pressures to test the applicability of the law of effective stresses to rocks. Experimental results show that this field still is wide open for further research.

Conclusion

Many of the details on the work on fracture are neglected in this review. However, it summarizes the state of recent research activities as well as the state of knowledge on rock fracture. Extensive experimental testing will be necessary and a large variety of new concept have to be considered to fully understand the process of fracture in rocks.

REFERENCES

[1]   W. Wawersik (1968): A detailed analysis of rock failure in laboratory compression tests. PhD thesis, Univ. of Minn.

[2]   N.G.W. Cook (1965): The failure of rock . Int. J. Rock Mech. Minn. Sci., Vol. 2,4.

[3]   B.P. Huges and Chapman (1966): The complete stress-strain curve for concrete in direct tension. Bull. Rilem. No. 30.

[4]   Z.T. Bieniawski (1967): Determination of rock properties. South Afr. CSIR Rep., No. Meg. 518

[5]   F. Rummel and Fairhurst (1970): Determination of the post-failure behaviour of brittle rock using a servo-controlled testing machine Rock Mech. 2, 189-204.

[6]   L. Fairhurst (1971): Fundamental considerations relating to the strength or rock. Coll. Fracture Mechanisms in Rocks, Bochum, 1971 Veröff. Inst. Bochum etc. Felsmech., Univ. Karlsruhe 1972, Heft 55.

[7]   J. Hudson et al. (1972): Soft, stiff and servo-controlled testing machines, a reviw. Eng. Geol., 6(3).

[8]   C. Fairhurst (1973): Estimation of the mechanical properties of rock masses. ARPA-Report, Univ. Minn.

[9]   K.V. Terzaghi (1929): Effect of minor geol. de ails on safety of dams. T.P. 215, ALME.

[10] K.V. Terzaghi (1945): Stress conditions for the failure of saturated concrete and rock. Proc. ASTM 45.

[11]   C.B. Raileigh and J. Handin (1972): Manmade earthquakes and earthquake
          control. Conf. Flow in Fiss. Rocks, Stuttgart.

[12]   Raileigh and Dieterich: pers. comm. (1972).

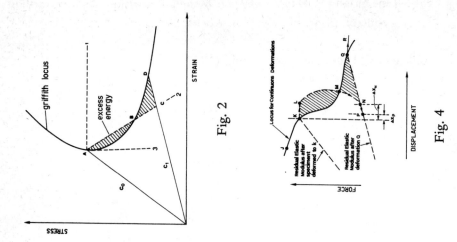

Fig. 2

Fig. 4

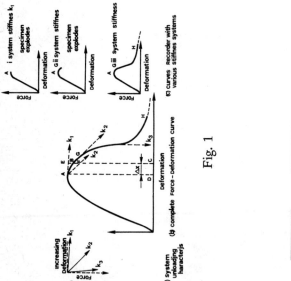

Fig. 1

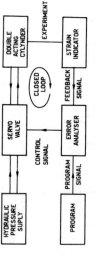

Fig. 3

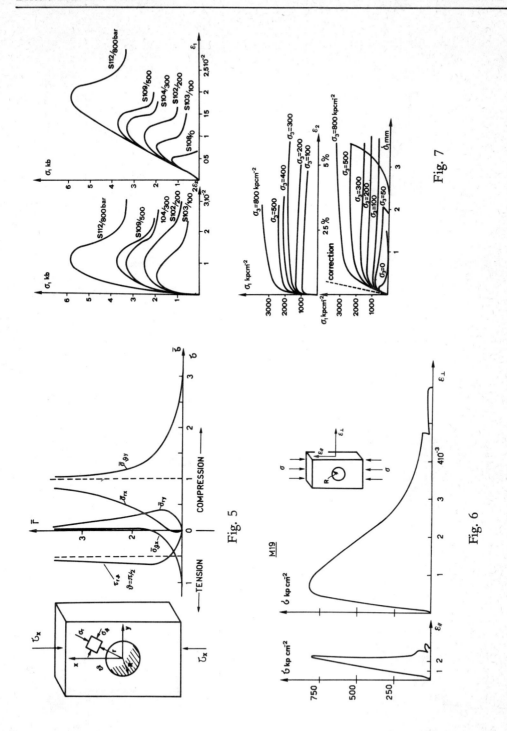

Fig. 5

Fig. 6

Fig. 7

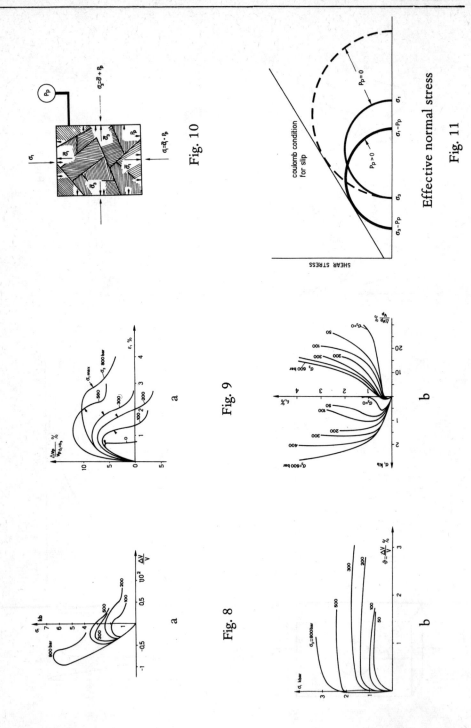

Fig. 10

Fig. 11

Effective normal stress

Fig. 9

Fig. 8

FAILURE MECHANISM OF JOINTED ROCK

H. K. KUTTER
Institut für Geologie
Ruhr-Universität Bochum

---

* All figures quoted in the text are at the end of the lecture.

Introduction

The structural geologist knew for a long time that the earth's crust is highly fractured, cracked and jointed and not, as the engineers for too long a time liked to assume, continuous, homogeneous, and isotropic and therefore the ideal solid material. This lack of appreciation of the geological facts was not purely the result of poor communication between the two disciplines — there were some practice-oriented geologists [4, 17] and fewer engineers [12, 14] who early recognized and advocated the necessity to take these structural discontinuities into consideration in the designs — but rather caused by the inability to deal analytically with such a complex material as the discontinuous medium.

The last decade, however, has brought considerable change in this attitude. A serious attempt is being made to explore the characteristics and the deformational and failure mechanism of jointed rock by means of systematic and controlled laboratory experiments on physical models and to develop therefrom simplified analytical models. This process is far from being completed and consequently a lecture on this subject cannot yet present a generalized theory but simply an up-to-date review of the past research findings.

Joints, faults and bedding planes form weak links in the composite material of a rock mass. In a one-dimensional structure the strength of the weakest link determines the strength of the entire system. In a two- or three-dimensional structure the failure of one element does not necessarily lead to rupture unless the weak element passes through the entire structure. But it alters the strength and deformational properties, causes local anisotropy and consequently a change in the stress distribution. The onset of local failure does therefore not mean collapse of the entire structure; it generally marks the transition from elastic to plastic and ductile behaviour.

Effect of Single Joint or Joint Set

The case of a single plane of weakness within a mass of intact material is a relatively simple one and its analytical solution is easy. A fault plane, a thin seam of shale, a thin weathered zone are geological examples for this type of discontinuity.

The strength of a material with a single plane of weakness is determined by the fact whether sliding along this plane can take place. Magnitude and orientation (with respect to the plane of weakness) of the principal stresses together with the shear strength parameters of the weak plane are the critical parameters which determine the onset of sliding. Starting from Coulomb's criterion for shear failure,

$$\tau = c + \sigma \tan \phi$$

where c is the cohesion and ϕ the angle of friction of the discontinuity, this can be rewritten in terms of the principal stresses

$$\sigma_1 - \sigma_3 = \frac{2c + 2 \sigma_3 \tan \phi}{(1 - \tan \phi \cot \beta) \sin 2\beta}$$

where β = angle between the plane of weakness and the minimum principal stress (Fig. 1).

Since the principal stress difference tends towards infinity as the angle β approaches $90°$ on one side and ϕ on the other, it follows that sliding failure is only possible when

$$\phi < \beta < \pi/2$$

The graphical representation of the above failure criterion (Fig. 2) shows further that the strength increases with σ_3 but reaches its minimum when the weak plane is inclined at the angle $\beta = 45° - \phi/2$. To complete the failure criterion also the strength of the solid material has to be taken into consideration. If the weak plane is unfavourably inclined failure can take place along another plane which passes through the intact material. Again, a Coulomb type failure criterion can be applied, only that this time the material constants c_o and ϕ_o are those of the intact material:

$$\tau = c_o + \sigma \tan \phi_o$$

The horizontal lines in Fig. 2, which represent the failure criterion for the intact material, intersect the respective parabola at two points. For those β values which fall between these points failure takes place by sliding along the plane of weakness, for inclinations outside these points the failure surface passes through the intact material.

The effect of a single plane of weakness on the mass behaviour becomes particularly clear when it is demostrated by means of the Mohr diagram.

Although the above results have been derived for a single plane of weakness only, they are equally valid for an entire set of parallel joints and for a homogeneous, but anisotropic material.

Effect of Multiple Intersecting Joint Sets

The above theory for a single set of joints can be extended and applied to a rock mass with several intersecting joint sets. Figure 3 shows how by superposition the composite failure criterion can be constructed. One can see, how the possibility of failure through the intact material becomes almost negligible for a rock mass with more than two symmetrical oriented joint sets.

This simple approach is unfortunately only of limited value, since it will only give the onset of failure in the sense that it indicates the stress combination at which the first sliding movement occurs. Other than in the case of the single joint, the problem is immediately complicated by the fact that a sliding movement at the intersection of two joints changes the continuity and consequently the shear resistance of one or both joint sets. This change of geometry alters the degree of anisotropy, the stress distribution and the strength properties. A different approach has therefore to be taken in order to obtain meaningful answers on progressive failure and the deformational behaviour of a rock mass.

As a first step it is possible to establish a wide band, with a definite upper and lower bound, within which the strength envelope of a jointed rock mass must lie. The lower bound is given by the failure envelope for sliding along a smooth joint, the upper bound by the Mohr failure envelope for the intact material (Fig. 4). In general one can say that at relatively high stresses the behaviour of the jointed rock will be closer to that of the intact rock, whereas at low stresses it will be determined predominantly by the properties of the joints. The more regular and continuous the jointing, the planer and smoother the joint surfaces and the smaller the degree of interlocking between the individual blocks, the nearer will be the failure envelope to the lower bound. Should there be such a thing as a rock mass with randomly oriented joints then its failure characteristics would be in the upper region of the defined failure band. The width of the band itself is best characterized by the ratio between shear strength of the intact material to the sliding resistance of the joints.

It seems possible that with a few fundamental considerations the width of the band can be narrowed down:

(1) Joints and discontinuities in the material greatly disturb the uniform distribution of stress since perfect contact across the continuity rarely exists. Partial and point contact leads to high stress concentrations and local crushing. Similarly as in the indirect tensile test, tensile fracturing of the blocks can occur if such point loads act on two opposite boundaries of a block. As a result

of this tensile failure mechanism, which is due to point contacts along the block boundaries, the overall strength of a jointed rock mass must be less than that of the continuum. The harder the rock, the rougher the joint surface (generated by tension rather than by shear), and the smaller the spacing of the joints, the higher will be the number of point contacts and consequently the more considerable is the reduction in strength. Even if the tensile fractures do not lead to a complete rupture of the blocks they alter their mechanical properties and introduce gross anisotropy.

(2) Pre-existing small cracks in the material increase in significance with respect to overall-strength with decreasing block size. The failure envelope therefore moves downward with increasing joint frequency.

(3) At the slightest sliding movement along one set of discontinuities an offset is introduced in all other intersecting joints. Their shear resistance is thus significantly altered due to the interlocking of blocks and it differs with direction. A build-up of anisotropy and dilatant shear are the result and the true failure envelope must therefore certainly lie somewhat higher than the lower bound.

(4) Rotation of the blocks causes interlocking and therefore an increase in shear strength. The smaller the blocks the more free they will be to rotate (at low normal stresses) and the higher will be the chances for tight interlocking. In the low stress region the strength of a jointed rock mass must therefore be higher than that of a single plane of weakness.

(5) In the case of intermittent discontinuous joints the lower bound is too pessimistic.

(6) The strength of a jointed rock mass is highly dependent on its previous load-deformation history. Rotation of blocks local crushing and dilation zones generated by any previous loading and deformation determine the present strength of the mass.

(7) If groundwater is present the water pressure has to be taken into consideration in the form of the effective stress concept when assessing the strength of the rock mass. Since dilation and compaction of the joints are common phenomena in the deformation process of a jointed rock, a sudden local rise or fall of water pressure has to be expected. The true strength envelope may therefore move back and forth between the upper and lower bound during a whole deformation process.

(8) The ratio between the size of structure (e.g. slope height) and the block size

determines the type of failure mechanism of a rock mass with multiple joint sets. If the joint spacing is of the same order of magnitude as the structure (slope, tunnel, foundation, etc.) then pure sliding along one or two discontinuities is the most likely failure mechanism; if the structure is two or more orders of mangitude larger than the joint spacing then the rock mass will behave similar to an anisotropic soil.

By applying these fundamental observations it should be possible to sketch for a specific situation an approximate but meaningful failure envelope between the limits of the upper and lower bounds. This simple procedure supplies an approximate design basis, but it gives no information as to the deformational properties of a jointed rock mass. Carefully designed experiments on physical models are therefore conducted by several researchers in order to gain a better understanding of the deformational **and** strength behaviour of a jointed rock mass.

Physical Models

In this section the results and conclusions of some recent model tests shall be summarized. No attempt is made to describe these investigations in full. Readers interested in more details are refered to the original publications.

The first experimental study of the strength and deformation behaviour of jointed rock masses was that by Müller and Pacher [13] Since then the modelling techniques have been refined and both model material and control of boundary conditions have been improved [2, 3, 6, 7, 9, 10, 11, 15].

a) Test on Plaster and Glass Blocks by Ergun [7]

The approach taken in this study was that of careful control of the boundary conditions, i.e. a uniform distribution of stress at the boundaries, in order to obtain true strength values for the jointed material. For this purpose a biaxial loading system was developed which applies uniform stresses to the boundaries of the model by means of closed box-shaped urethane rubber seals and Teflon-lined steel loading pads. The model consisted of cubical blocks of high-strength plaster and glass. The latter was used in order to test the stress distribution by photoelasticity (Fig. 5). Despite the very smooth joint surfaces high local stress concentrations were observed even before the onset of slipping and crushing. The models has one or two orthogonal joint sets which were either horizontal, vertical or inclined.

The experiments were carried out to compare the stress distribution and strength of a continuous model with those of discontinuous models, to study the stability, stress distribution and displacements around underground openings in discontinuous material, and to investigate the influence of the stress history of the excavation of a square tunnel in a rock mass with two orthogonal joint sets.

The results showed that the stress distribution in a jointed rock mass is greatly affected by

(i)   the inclination of the joints to the direction of the applied principal stresses
(ii)   the ratio of the applied stresses
(iii) the presence of voids along joints
(iv)   the stress history.

The convincing demostration of the dependency of the strength on the previous loading paths and cyles through which the material went is the main merit of this investigation. To simply assume that the strength of a zone around a tunnel which must have gone through complex loading and unloading cycles due to the excavation process, is the same as the strength of an undisturbed jointed rock mass is therefore incorrect and can lead to wrong design criteria.

b) Strength Tests on Plaster Models with Intermittent Joints by Brown [3]

In a series of triaxial compression tests on model rock masses the effect of intermittent joints was studied. With the introduction of so called rock bridges which interrupt the continuity of the joint, (they are frequently observed in the field) an additional variable complicates the interpretation of the results, but the model certainly becomes more realistic and new failure modes are discovered. Parallelepipedal and hexagonal blocks of high-strength gypsum plaster were put together into five different model configurations (Fig. 6) with a two-dimensional extent of jointing of 50, respectively 33 percent. These block systems were tested to failure under a range of constant confining pressures ($\sigma_2 = \sigma_3$).

The failure envelopes (drawn from the peak stresses obtained in these model tests) are clearly curved and appear to pass through the origin. They are best described by a parabolic rather than a linear equation:

$$\tau = \tau_o + Z'\sigma_n^\xi \qquad \text{or} \qquad \frac{\tau - \tau_o}{\sigma_c} = Z\left(\frac{\sigma_n}{\sigma_c}\right)^\xi$$

where τ_o, Z, Z and ξ are material constants and σ_c the compressive uniaxial strength. ξ varied for these experiments between 0.57 and 0.73, whereas ξ for an intact continuous block was 0.50. The curvature of the failure envelope clearly increasing with increasing degree of interlocking and decreasing extent of jointing (Fig. 7).

Significant features of the stress-strain curves were (1) development of stick-slip oscillations for all but the uniaxial tests, (2) post-peak sliding on failure planes with the development of some residual strength and (3) apparent increase in deformation modulus with an increase of confining pressure.

Seven different macroscopic failure modes (Fig. 8) could be observed at the completion of the tests (initial failure can have been a different one!):

A) Axial cleavage at low confining pressure
B) Shear failure along an approximately planar surface independent of joints
C) Collapse at low confining pressures as a result of block movement involving the opening of joints and delation of the simple.
D) Formation of a single composite shear plane, partly through intact material and partly along joints.
E) The formation of complex, non-planar shear failure surfaces, partly through intact material and partly along joints.
F) The formation of multiple conjugate shear planes through intact material at high confining pressures.
G) The formation of multiple conjugate shear planes partly through plaster and partly along joints.

The most significant conclusion which can be drawn from these test results is that in addition to sliding along joints and to shearing through solid material two other important failure modes, namely axial cleavage (A) and dilatational (C) failure take place in a jointed rock mass. The earlier discussed, simple concept of shear failure only is therefore of limited value.

c) **Examination of Strength and Deformational Behaviour of Single and Multiple Jointed Plaster Models by Einstein and Hirschfeld [6]**

A series of prismatic jointed plaster specimens (Fig. 9) were loaded triaxially up to failure. The test results are described in terms of strength, deformability and failure mode and can be summarized as follows:

1) Models with single joint and completely intact: The transition from sliding to fracturing of intact material occurs at the same critical normal stress as the transition from brittle to ductile behaviour and the transition from failure along a single joint to failure along multiple joints.

2) Multiple jointed models
   a) with no sliding along joints (joint parallel and/or normal to σ_1):

The strength increases and the deformability decreases with increasing joint-spacing, from orthogonally jointed to horizontally jointed to vertically jointed model configurations. b) with sliding along joints possible (joints at 45° angle to direction of σ_1):

At the low stress range failure occurs by sliding along pre-existing joints, at the higher range failure by fracture through intact material, frequently development of a number of fracture surfaces.

Based on these test results a generalized theory of fracture of jointed rock was developed (Fig. 10). In zone A and B sliding takes place along pre-existing joints if they are favourably inclined and consequently the Mohr envelope for smooth joints indicates failure of the jointed mass. If the joints are not favourably inclined the fracture passes through intact material and intersects the joints. The failure line lies in zone A between the envelopes for intact material and smooth joint, in zone B parallel to and about 5 to 10 per cent lower than the intact envelope. The transition from brittle to ductile behaviour takes place in zone B. Failure in zone C is only ductile and independent of pre-existing joints. An approximate failure envelope can thus be constructed for a jointed rock mass.

d) Strength Tests on an Interlocked Low-porosity Aggregate by Rosengren and Jaeger [16]

By heating coarse grained marble to around 600° C and thus generating almost complete separation along the grain boundaries, an excellent model for simulating randomly jointed rock was obtained. The mechanical properties of this material were examined in triaxial tests and it was found that

1) They were very different from those of soils.
2) The Mohr envelope (Fig. 11) passes through the origin, is very steep at low normal stresses, but its slope decreases very rapidly with increasing normal stress. At the higher stress range the strength increases to over 80 per cent of that of the intact rock.
3) Young's modulus increases with confining pressure, but only to about 30 per cent of the intact rock.
4) Permeability decreases rapidly with confining pressure and slightly with uniaxial stress.
5) The type of failure was shear fractures on a single plane at lower confining pressures, changing into a pattern of conjugate shears for greater confining pressures.

It is significant that even this tightly interlocked material has practically no shear strength at zero normal stress.

Concluding Remarks

In retrospect it has become obvious that neither the theoretical considerations nor the latest results and conclusions from model experiments are yet sufficient for the establishment of a general theory for the mechanical behaviour of a jointed rock mass, which could be directly applied in design practice. But the knowledge we have obtained so far on this subject allows us the prediction of an approximate range within which the strength envelope will lie.

There are and will be attempts to narrow this range even further. No mention was made here of the promising work with analytical block models which uses either the powerful finite element method [8] or the dynamic relaxation technique [5]. So far these numerical methods are still dealing with a continuum or with an extremely simplified discontinuum. More refined physical models will equally help in narrowing the band: The smooth-walled joints between plaster bricks which are used in most models have little dilation under shear and hardly the rough, matching surfaces of tensile fractures in real rock. Barton [1] in his slope models has for the first time developed a modelling technique where these essential features of rock joints have been integrated. Joints sets of different frictional properties are another aspect which so far was hardly considered.

Finally, when talking about mechanics and failure of jointed rock, also the so far practically untouched problem of dynamic failure (e.g. blasting, earthquake damage, etc.) of rock masses should be at least mentioned. The study of the transmission and dispersion of waves in jointed rock would be one aspect which should be dealt with in future research.

REFERENCES

[1]   Barton, N.R., "A model study of rock-joint deformation", Int. Journal of Rock Mechanics and Mining Sciences, Vol. 9, No. 5, pp. 579-602.

[2]   Brown, E.T. and Trollpe, D.H., "Strength of a model of jointed rock" Journal of the Soil Mechanics and Foundation Division, ASCE, Vol. 96, No. SM2, 1970, pp. 685-704.

[3]   Brown, E.T., "Strength of models of rock with intermittent joints", Journal of the Soil Mechanics and Foundation Division, ASCE, Vol. 96, No. SM6, 1970, pp. 1935-1949.

[4]   Clar, E., "Zur Darstellung der Klüftigkeit von Felsaufschlüssen", Geologie und Bauwesen, Vol. 11, No. 1, 1939, p. 1.

[5]   Cundall, P.A., "A computer model for simulating progressive, large-scale movements in blocky rock systems", Proceedings of ISRM Symposium on 'Rock Fracture', Nancy 1971, Paper II-8.

[6]   Einstein, H.H. and Hirschfeld, R.C., "Model studies on mechanics of jointed rock", Journal of the Soil Mechanics and Foundation Division, ASCE, Vol. 99, No. SM3, 1973, pp. 229-248.

[7]   Ergun, I., "Stress distribution in jointed media", Proceedings of the 2nd Congress of the Int. Soc. Rock Mechanics, Vol. 1, Paper No. 2-31, Belgrade, 1970.

[8]   Goodman, R.E., Taylor, E.L. and Brekke, T.L., "A model for the mechanics of jointed rock", Journal of the Soil Mechanics and Foundation Division, ASCE, Vol. 94, No. SM3, 1968, pp. 637-659.

[9]   Hayashi, M., "Strength and dilatancy of a jointed mass", Proceedings of the 1st Congress of the Int. Soc. Rock Mehcanics, Vol. 1, Lisbon 1966, pp. 295-301.

[10]   John, K.W., "Engineering methods to determine strength and deformability of regularly jointed rock", Proceedings, 11th Symposium on Rock Mechanics, Berkeley, Calif. 1969, AIME 1970.

[11]   Ladanyi, B. and Archambault, G., "Simulation of shear behaviour of a jointed rock mass", Proceedings, 11th Symposium on Rock Mechanics, Berkeley, Calif. 1969, AIME 1970, pp. 105-125.

[12]   Müller, L., "Der Kluftkörper", Geologie und Bauwesen, Vol. 18, No. 1, 1950, p. 57.

[13]   Müller, L. and Pacher, F., "Modellversuche zur Klärung der Bruchgefahr geklüfteter Medien", Felsmechanik und Ingenieurgeologie, Supplement II, 1965, pp. 7-24.

[14]   Pacher, F., "Kennziffern des Flächengefüges", Geologie und Bauwesen, Vol. 24, No. 3/4, 1959, p. 223.

[15]   Rosenblad, J.L., "Failure modes of models of jointed rock masses", Proceedings, 2nd Congress of the Int. Soc. Rock Mechanics, Vol. 1, Paper 3-11, Belgrade, 1970.

[16]   Rosengren, K.J. and Jaeger, J.C., "The mechanical properties of an interlocked low-porosity aggregate", Géotechnique, Vol. 18, 1968, pp. 317-326.

[17]   Stini, J., "Technische Gesteinskunde", Springer, Wien, 1929.

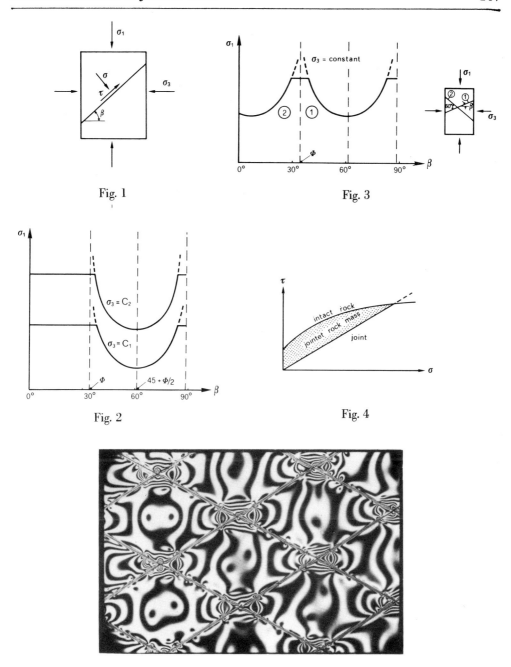

Fig. 1

Fig. 3

Fig. 2

Fig. 4

Fig. 5

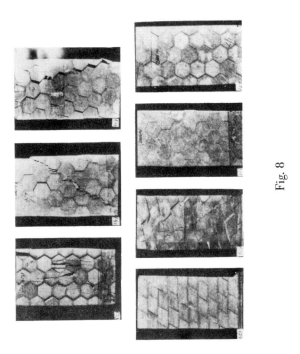

Fig. 8

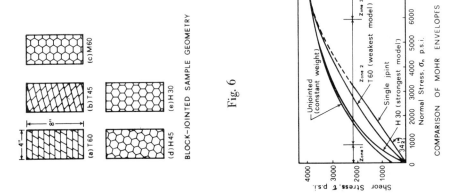

BLOCK-JOINTED SAMPLE GEOMETRY

(a) T 60   (b) T 45   (c) M 60

(d) H 45   (e) H 30

Fig. 6

COMPARISON OF MOHR ENVELOPES

Fig. 7

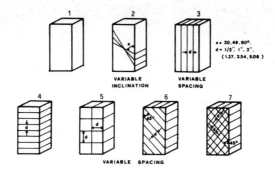

Joint Configuration and Testing Sequence of Model Rock

Fig. 9

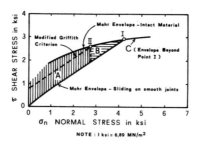

NOTE : I ksi = 6,89 MN/m²

Behavior of Jointed Model Rock Characteristic Boundaries and Zones

Fig. 10

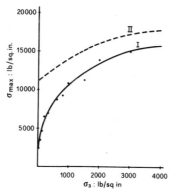

Curve I: maximum differential stress attained.
σ_{max} plotted against confining pressure
σ_3 for 'granulated' marble. Curve II : corresponding average curve for the original marble

Fig. 11

INTRODUCTORY LECTURE
ON FINITE ELEMENT ANALYSIS
FOR JOINTED ROCKS

R. E. GOODMAN
Associate Professor
University of California, Berkeley
Department of Civil Engineering

---

* All figures quoted in the text are at the end of the lecture.

Introduction

Finite element analysis is an application of the **direct stiffness method** of structural analysis; the steps of the method are as follows:

1. Conceive of **a prototype** embracing the important feature of actual structure. Subdivide it into **nodal points** and **elements**.
2. Form the matrix connecting displacements and forces at each nodal point of element (element stiffness matrices).
3. Assemble the structural stiffness matrix from each element stiffness matrix.
4. Introduce applied forces and write the equations of equilibrium.
5. Introduce displacement boundary conditions.
6. Solve for displacements at nodal points.
7. From known displacement in each nodal point of an element, find element stresses.
8. Recycle for non-linear properties

To illustrate, Example 1 solves a simple one dimensional problem.

EXAMPLE 1

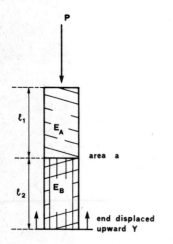

Calculate the stress and displacements in the simple structure.

Step 1

2 *elements*

3 *nodal points*

1

A $k_A = aE_A / \ell_1$

2

B $k_B = aE_B / \ell_2$

3

Step 2

element A stiffness matrix:

J

A

$$\left\{ \begin{matrix} F_I \\ F_J \end{matrix} \right\} = \frac{aE_1}{\ell_1} \begin{bmatrix} 1 & -1 \\ -1 & 1 \end{bmatrix} \left\{ \begin{matrix} u_I \\ u_J \end{matrix} \right\}$$

I

extension positive

or

$$\{F\}_A = \begin{bmatrix} k_A & -k_A \\ -k_A & k_A \end{bmatrix} \{u\}_A$$

$$[K_A]$$

J

B similarly $$[K_B] = \begin{bmatrix} k_B & -k_B \\ -k_B & k_B \end{bmatrix}$$

I

Step 3

structural stiffness matrix:

$$
\begin{Bmatrix} F_1 \\ F_2 \\ F_3 \end{Bmatrix} = \underbrace{\begin{bmatrix} k_A & -k_A & 0 \\ -k_A & (k_A + k_B) & -k_B \\ 0 & -k_B & k_B \end{bmatrix}}_{[K]} \begin{Bmatrix} u_1 \\ u_2 \\ u_3 \end{Bmatrix}
$$

Step 4

the applied forces $= \begin{Bmatrix} -P \\ 0 \\ X \end{Bmatrix}$ where X stands for the unknown force of reaction at node 3.

the external forces on the elements due to thir deformations are

$$[K]\{u\}$$

Equilibrium requires:

$$
\begin{Bmatrix} -P \\ 0 \\ X \end{Bmatrix} = [K] \begin{Bmatrix} u_1 \\ u_2 \\ u_3 \end{Bmatrix}
$$

Step 5

the displacement $u_3 = Y$

$$
\begin{Bmatrix} -P \\ 0 \\ X \end{Bmatrix} = \begin{bmatrix} k_A & -k_A & 0 \\ -k_A & (k_A + k_B) & -k_B \\ 0 & -k_B & k_B \end{bmatrix} \begin{Bmatrix} u_1 \\ u_2 \\ Y \end{Bmatrix} \text{ (Known)}
$$

Remove knowns from right side, and identify equations to be solved:

$$\begin{Bmatrix} -P \\ 0 + k_B Y \end{Bmatrix} = \begin{bmatrix} k_A & -k_A \\ -k_A & k_A + k_B \end{bmatrix} \begin{Bmatrix} u_1 \\ u_2 \end{Bmatrix}$$

and

$$\{X - k_B Y\} = - k_B u_2$$

Step 6

   inverting the first two equations yields

$$u_1 = y - \frac{P}{k_A} - \frac{P}{k_B}$$

and

$$u_2 = y - \frac{P}{k_B}$$

which together with the condition $u_3 = Y$ completely determines the displacements

Step 7

$$\sigma_A = \frac{(F_I)_A}{a} = \frac{1}{a}(k_A - k_A)\begin{Bmatrix} u_1 \\ u_2 \end{Bmatrix} = \frac{-P}{a}$$

with the same end result for B.

Element Stiffness

   The steps are as above in an actual finite element analysis but the elements are two or three dimensional, as required, and the solution may be recycled to pursue stress dependent properties. The simplest element, and one of the original standard elements, is a triangle defined by three nodal points and derived assuming a constant state of strain throughout.

   The element stiffness matrix is (6 x 6) since

$$\{F\} = [K]\{u\} \tag{1}$$

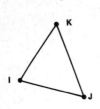

and $\{F\}$ and $\{u\}$ are lists of six numbers

$$\{F\} = \left(F_{x_I} F_{y_I} F_{x_J} F_{y_J} F_{x_K} F_{y_K} \right)^T$$

and

$$\{u\} = \left(u_{x_I} u_{y_I} u_{x_J} u_{y_J} u_{x_K} u_{y_K} \right)^T$$

Let strains $(\epsilon_x \epsilon_y \gamma_{xy})^T$ be related to corner displacements (nodal displacements) $\{u\}$ by $\{\epsilon\} = [L_o]\{u\}$; then the element stiffness matrix

(2)
$$[K] = a [L_o]^T [C][L_o]$$

where $[C]$ is the stress strain relationship $\{\sigma\} = [C]\{\epsilon\}$.

Setting the local coordinate origin at I

$$(ie, \ x_I = y_I = 0):$$

(1) a, the area of the triangle, is numerically equal to $1/2 \ (x_J y_K - x_K y_J)$.

For small strains, with the local origin at node I:

(2)
$$[L_o] = \frac{1}{2a} \begin{pmatrix} y_J - y_K & 0 & y_K & 0 & -y_J & 0 \\ 0 & x_K - x_J & 0 & -x_K & 0 & x_J \\ x_K - x_J & y_J - y_K & -x_K & y_K & x_J & -y_J \end{pmatrix}$$

For isotropic rock in plane strain:

(3)
$$[C] = \frac{E(1-\nu)}{(1+\nu)(1-2\nu)} \begin{pmatrix} 1 & \nu/(1-\nu) & 0 \\ \nu/(1-\nu) & 1 & 0 \\ 0 & 0 & \frac{1-2\nu}{2(1-\nu)} \end{pmatrix}$$

where E is the Young's Modulus and ν is Poisson's ratio. The complete derivation of these relationships and other stress strain relationships for anisotropic rocks are given in a monograph in press. (*)

(*) Goodman, R.E., "Methods of Geological Engineering in Discontinuous Rock", (West Publishing Company).

Example 2 illustrates computation of the stiffness matrix for a constant strain triangle.

EXAMPLE 2

Find the stiffness matrix for the triangular finite element
Rock is isotropic with

$$\nu = 0.2$$

$$E = 1000$$

Initial stresses are

$$\sigma_{x,o} = -10$$

$$\sigma_{y,o} = -5$$

$$\tau_{xy,o} = -1$$

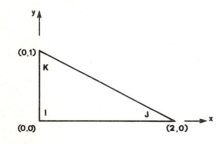

The strain displacement matrix is

$$(L_o) = \frac{1}{2} \begin{bmatrix} -1 & 0 & 1 & 0 & 0 & 0 \\ 0 & -2 & 0 & 0 & 0 & 2 \\ -2 & -1 & 0 & 1 & 2 & 0 \end{bmatrix}$$

The stress strain matrix is

$$(C) = \begin{bmatrix} 1111 & 278 & 0 \\ 278 & 1111 & 0 \\ 0 & 0 & 417 \end{bmatrix}$$

Thus, the stiffness matrix (K) is, by (2)

$$(K) = \frac{1}{4} \begin{bmatrix} -1 & 0 & -2 \\ 0 & -2 & -1 \\ 1 & 0 & 0 \\ 0 & 0 & 1 \\ 0 & 0 & 2 \\ 0 & 2 & 0 \end{bmatrix} \begin{bmatrix} 1111 & 278 & 0 \\ 278 & 1111 & 0 \\ 0 & 0 & 417 \end{bmatrix} \begin{bmatrix} -1 & 0 & 1 & 0 & 0 & 0 \\ 0 & -2 & 0 & 0 & 0 & 2 \\ -2 & -1 & 0 & 1 & 2 & 0 \end{bmatrix}$$

giving

$$
(K) = \quad I \qquad\qquad J \qquad\qquad K
$$

$$
\begin{array}{c}
I \\[20pt]
J \\[20pt]
K
\end{array}
\left[
\begin{array}{cc cc cc}
694 & 347 & -278 & -208 & -416 & -139 \\
347 & 1215 & -139 & -104 & -208 & -1111 \\[10pt]
-278 & -139 & 278 & 0 & 0 & 139 \\
-208 & -104 & 0 & 104 & 208 & 0 \\[10pt]
-416 & -208 & 0 & 208 & 416 & 0 \\
-139 & -1111 & 139 & 0 & 0 & 1111
\end{array}
\right]
$$

Element Stresses

The stresses in a constant strain triangle are related to the displacements of the corners (nodal point displacements) by:

(3)
$$
\begin{pmatrix} \sigma_x \\ \sigma_y \\ \tau_{xy} \end{pmatrix} = [C][L_o]\{u\}
$$

Discontinuities

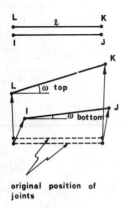

original position of joints

Jointed rock masses are modelled by solid elements linked by special "joint elements" consisting of two lines of two nodal points.

The "strain vector" for a joint element $\{\epsilon_J\}$ is defined by the relative displacements and rotations of the two walls as measured at the joint center.

$$
\{\epsilon_J\} = (\Delta u_o, \Delta v_o, \Delta \omega_o)^T
$$

The "strains" are related to nodal displacements by

$$
\begin{pmatrix} \Delta u_o \\ \Delta v_o \\ \Delta \omega_o \end{pmatrix} = \begin{pmatrix} -\dfrac{1}{2} & 0 & -\dfrac{1}{2} & 0 & \dfrac{1}{2} & 0 & \dfrac{1}{2} & 0 \\ 0 & -\dfrac{1}{2} & 0 & -\dfrac{1}{2} & 0 & \dfrac{1}{2} & 0 & \dfrac{1}{2} \\ 0 & \dfrac{1}{\ell} & 0 & -\dfrac{1}{\ell} & 0 & \dfrac{1}{\ell} & 0 & -\dfrac{1}{\ell} \end{pmatrix} \begin{Bmatrix} u_I \\ v_I \\ u_J \\ v_J \\ u_K \\ v_K \\ u_L \\ v_L \end{Bmatrix} \quad (4)
$$

The local joint stresses are $\{\sigma\}_{sn} = (\tau_{sn} \ \sigma_n M_o)^T$ and

$$
\begin{pmatrix} \tau_{sn} \\ \sigma_n \\ M_o \end{pmatrix} = \begin{pmatrix} k_s & 0 & 0 \\ 0 & k_n & 0 \\ 0 & 0 & k_\omega \end{pmatrix} \begin{pmatrix} \Delta u_o \\ \Delta v_o \\ \Delta \omega_o \end{pmatrix} \quad (5)
$$

k_ω can be evaluated by considering the moment and rotation when nodes I and J are fixed and all the force is applied to either node K or L. This yields

$$
k_\omega = \frac{\ell^3 k_n}{4} \quad (6)
$$

Nodal point forces are related to $\{\sigma\}_{sn}$ by

$$(7) \quad
\begin{Bmatrix}
F_{s_I} \\
F_{n_I} \\
F_{s_J} \\
F_{n_J} \\
F_{s_k} \\
F_{m_k} \\
F_{s_L} \\
F_{n_L}
\end{Bmatrix}
=
\begin{pmatrix}
-\dfrac{\ell}{2} & 0 & 0 \\[6pt]
0 & -\dfrac{\ell}{2} & \dfrac{1}{\ell} \\[6pt]
-\dfrac{\ell}{2} & 0 & 0 \\[6pt]
0 & -\dfrac{\ell}{2} & -\dfrac{1}{\ell} \\[6pt]
+\dfrac{\ell}{2} & 0 & 0 \\[6pt]
0 & +\dfrac{\ell}{2} & \dfrac{1}{\ell} \\[6pt]
+\dfrac{\ell}{2} & 0 & 0 \\[6pt]
0 & +\dfrac{\ell}{2} & -\dfrac{1}{\ell}
\end{pmatrix}
\begin{pmatrix}
\tau_{sn} \\
\sigma_n \\
M_o
\end{pmatrix}$$

The local joint element stiffness relates $\{F\}_{sn}$ to $\{u\}$ and is therefore

$$(8) \quad [K_{sn}] = \frac{\ell}{4}
\begin{pmatrix}
k_s & 0 & k_s & 0 & -k_s & 0 & -k_s & 0 \\
0 & 2k_n & 0 & 0 & 0 & 0 & 0 & -2k_n \\
k_s & 0 & k_s & 0 & -k_s & 0 & -k_s & 0 \\
0 & 0 & 0 & 2k_n & 0 & -2k_n & 0 & 0 \\
-k_s & 0 & -k_s & 0 & k_s & 0 & k_s & 0 \\
0 & 0 & 0 & -2k_n & 0 & 2k_n & 0 & 0 \\
-k_s & 0 & -k_s & 0 & k_s & 0 & k_s & 0 \\
0 & -2k_n & 0 & 0 & 0 & 0 & 0 & 2k_n
\end{pmatrix}$$

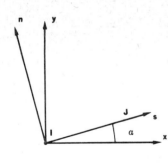

Of course this must be rotated to find the term by term contributions to the structural stiffness matrix with respect to global x y coordinates.

Let $[K_{IJ}]_{sn}$ be the 2 x 2 submatrix of stiffness coefficients relating forces in the s and n directions at node I to displacement in the s and n directions at J. Then the rotated stiffness submatrix for the joint element is

$$[K_{IJ}]_{xy} = [T_I]^T [K_{IJ}]_{sn} [T_I]   \qquad (9)$$

where $[T_I]$ is the rotation matrix

$$[T_I] = \begin{bmatrix} \cos\alpha & \sin\alpha \\ -\sin\alpha & \cos\alpha \end{bmatrix}$$

Assembly of the Structural Equations

This is a "bookkeeping" problem and is best discussed through another example. We will assemble the 2n x 2n stiffness matrix for a simple structure with n = 6 nodal points.

EXAMPLE 3

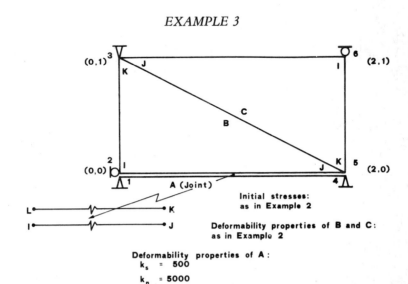

Initial stresses:
as in Example 2

Deformability properties of B and C:
as in Example 2

Deformability properties of A :
$k_s = 500$
$k_n = 5000$

Initial stresses: as in Example 2

Deformability properties of B *and* C: *as in Example 2*

Deformability properties of A:

$$k_s = 500$$

$$k_n = 5000$$

Elements B and C are identical to the element previously calculated in Example 2 if we name the corners identically. The joint, element A, has stiffness terms as given by equation 8.

The 12 x 12 symmetric stiffness matrix for the structure of Example 3 is

$$
\begin{pmatrix}
k_{II}^{A} & k_{IL}^{A} & 0 & k_{IJ}^{A} & k_{IK}^{A} & 0 \\
k_{LI}^{A} & k_{LL}^{A} + k_{II}^{B} & k_{IK}^{B} & k_{LJ}^{A} & k_{LK}^{A} + k_{IJ}^{B} & 0 \\
0 & k_{KI}^{B} & k_{KK}^{B} + k_{JJ}^{C} & 0 & k_{KJ}^{B} + k_{JK}^{C} & k_{JI}^{C} \\
k_{JI}^{A} & k_{JL}^{A} & 0 & k_{JJ}^{A} & k_{JK}^{A} & 0 \\
k_{KI}^{A} & k_{KL}^{A} + k_{JI}^{B} & k_{JK}^{B} + k_{KJ}^{C} & k_{KJ}^{A} & k_{KK}^{A} + k_{JJ}^{B} + k_{KK}^{C} & k_{KI}^{C} \\
0 & 0 & k_{IJ}^{C} & 0 & k_{IK}^{C} & k_{II}^{C}
\end{pmatrix}
$$

Substituting the required stiffness terms gives (K) =

1	2	3	4	5	6	7	8	9	10	11	12
+250	0	-250	0	0	0	+250	0	-250	0	0	0
0	+5000	0	-5000	0	0	0	0	0	0	0	0
-250	0	935	342	-407	-139	-250	0	-28	-204	0	0
0	-5000	342	6213	-204	-1111	0	0	-139	-102	0	0
0	0	-407	-204	685	0	0	0	0	342	-278	-139
0	0	-139	-1111	0	1213	0	0	342	0	-204	-102
+250	0	-250	0	0	0	+250	0	-250	0	0	0
0	0	0	0	0	0	0	+5000	0	-5000	0	0
-250	0	-28	-139	0	342	-250	0	935	0	-407	-204
0	0	-204	-102	342	0	0	-5000	0	6213	-139	-1111
0	0	0	0	-278	-204	0	0	-407	-139	684	342
0	0	0	0	-139	-102	0	0	-204	-1111	342	1213

"Applied" Loads — The Structural Load Vector

Jointed rock structures may be loaded by gravity, pseudo static acceleration, directly applied point forces or pressures, and water forces.

The load vector due to any of these causes will be called $\{F_e\}$; it is made up of contributions from each node of each triangle. Forces $\{F_o\}$ are transferred at the nodes to maintain the initial stresses so that the net applied external force is

$$(10) \qquad\qquad \{F_e\} - \{F_o\} = \{F_{net}\}$$

Residual Stress Forces in each constant strain triangle are:

$$(11) \qquad\qquad \{F_o\} = a(L_o^T)\{\sigma_o\}$$

where $\{\sigma_o\}$ defines the initial stress state in the element.

$$\{\sigma_o\} = \left(\sigma_{x_o}, \sigma_{y_o}, \tau_{xy,o}\right)^T$$

and a is the area of the triangle $\frac{1}{2}(x_J y_K - x_K y_J)$. If there is no initial moment in a joint, its x,y initial forces are related to x,y initial stresses by

$$(12) \qquad\qquad \{F_o\}_J = [T]^T (B) (T_{o.J})\{\sigma_o\}_{x.y}$$

where

$$T = \begin{bmatrix} (T_1) & 0 & 0 & 0 \\ 0 & (T_1) & 0 & 0 \\ 0 & 0 & (T_1) & 0 \\ 0 & 0 & 0 & (T_1) \end{bmatrix}$$

$$
B = \begin{pmatrix}
-\dfrac{\ell}{2} & 0 \\[2mm]
0 & -\dfrac{\ell}{2} \\[2mm]
-\dfrac{\ell}{2} & 0 \\[2mm]
0 & -\dfrac{\ell}{2} \\[2mm]
+\dfrac{\ell}{2} & 0 \\[2mm]
0 & +\dfrac{\ell}{2} \\[2mm]
+\dfrac{\ell}{2} & 0 \\[2mm]
0 & +\dfrac{\ell}{2}
\end{pmatrix}
$$

and

$$
T_{\sigma,J} = \begin{pmatrix}
\dfrac{-\sin 2\alpha}{2} & \dfrac{\sin 2\alpha}{2} & \cos 2\alpha \\[2mm]
\sin 2\alpha & \cos 2\alpha & -2\sin 2\alpha
\end{pmatrix}
$$

Gravity and Inertia Forces. The external forces at each node of a constant strain triangle due to a body force are 1/3 of the total body force applied to the whole triangle. Let x be horizontal and y positive upwards. The unit weight of the rock is γ (e.g. 0.02MN/m^3) and in addition to gravity, g, the element is constantly accelerated at a rate of C g in a direction producing an inertia force α degrees counterclockwise from x. Then the external forces at each node are

$$
\begin{pmatrix} F_x \\ F_y \end{pmatrix} = \frac{\gamma a}{3} \begin{pmatrix} C \cos \alpha \\ -1 + C \sin \alpha \end{pmatrix} \tag{13}
$$

where a is the area of the triangle.

Water Forces. If a triangular element is saturated and subject to potential gradients $\partial h/\partial x$ and $\partial h/\partial y$, the additional body force at each node due to the action of the water is

(14)
$$\begin{pmatrix} F_x \\ F_y \end{pmatrix} = \frac{\gamma_w a}{3} \begin{pmatrix} -\partial h/\partial x \\ 1 - \partial h/\partial y \end{pmatrix}$$

where γ_w is the unit weight of the water (e.g. $0.01 MN/m^3$).

Directly Applied Line Loads can of course, be added to the load vector $\{F_e\}$ at the appropriate nodal points. Rock bolt forces, though approximately point loads, may be input this way. To model bolts properly, however, one must account for the additional stiffness contributions of the bolts in shear and in extension.

EXAMPLE 4

Find the load vector for the structure of Example 3. The only contribution to the net applied load comes from residual stress. Applying (11) for element B gives

$$\{F_o\}_B = (6, \; 5.5, \; -5, \; -0.5, \; -1, \; -5)^T$$

Because triangle C is obtainable from triangle B by rotation through $180°$, its vector of initial load $\{F_o\}_C = - \{F_o\}_B$.
The initial loads on the joint to maintain the initial stresses are

$$\{F_o\}_C = (1, \; 5, \; 1, \; 5, \; -1, \; -5, \; -1, \; -5)^T$$

Summing all elements contributing to the structure gives the net load vector

$$\{F\} = - \{F_o\} = \begin{matrix} & \overset{1}{x} \quad y & \overset{2}{x} \quad y & \overset{3}{x} \quad y & \overset{4}{x} \quad y & \overset{5}{x} \quad y & \overset{6}{x} \quad y \\ & (-1,-5, & -5,-0.5, & -4,+4.5, & -1,-5, & +5,0.5, & 6,5.5) \end{matrix}^T$$

Iterative Solution to Simulate Real Properties of Joints

As discussed by Goodman and Dubois (1973) either properties (stiffnesses) or initial loads may be varied successively to force the results into an "acceptable" combination; joints are highly non linear. **Figure 1** shows the displacements Δu_0 and Δv_0 as a function of σ and τ. **Figure 2** shows how varying the initial load can force the solution onto the constitutive curve for joint closing by decompression of a prestressed block. Goodman (*), and Goodman and St. John (1974) discuss this in detail and present a small computer program in which hyperbolic compression behavior, peak and residual shear strength behavior, and dilatancy are modeled. The peak shear strength variation with normal stress is expressed by the formula of Ladanyi and Archambault (1969) with a continuous variation of contact area and dilatancy according to a power of normal stress. Iterations are programmed by load transfer but variable stiffness cycles are computed each time the program is restarted. In this program, the joint normal stiffness is computed from the initial stress and only the shear stiffness is input. The input information is restricted by making many assumptions within the body of the program. This helps a "user" to avoid incompatibilities in input properties. The basic problem is that while convergent output suggests a stable structure, divergent results may reflect either true structural instability, or mathematical problems.

EXAMPLE 5

   Consider a block between two joints loaded by a shear couple as shown in the Figure. Rotation first closes the joint beyond the permissible maximum closure V_{mc}. Subsequent iterations produce a convergent final state of displacement and stress as shown in Figure 3.

Conclusion

   Finite element computations enable non-linear, discontinuous, anisotropic, path-dependent, and otherwise non "ideal" properties to be considered in rock mechanics. The flow of ideas in such computations has been briefly traced and the stiffness matrix for a new joint element has been presented. References listed below may be consulted for further information.

(*) Monograph in preparation previously cited.

REFERENCES

[1]   Clough, R.N., "The Finite Element Method in Plane Stress Analysis,"
       Proceedings 2nd ASCE Conference on Electronic Computation,
       1960, pp. 345-378.

[2]   Ghaboussi, J., E. Wilson, and J. Isenberg, "Finite Element Analysis for Rock
       Joints and Interfaces," Journal Soil Mechanics and Foundations
       Division, Proceedings ASCE, 1973, Vol. 99, No. SM 10, pp. 833-848.

[3]   Goodman, R.E., "The Mechanical Properties of Joints," Proceedings 3rd.
       Congress of the International Society of Rock Mechanics, 1973, Vol.
       1, Part 2, Supplementary Reports Volume (pre Congress volume).

[4]   Goodman, R.E.,"Methods of Geological Engineering in Discontinuous Rocks,"
       West Publishing Company, in preparation (estimated 1975).

[5]   Goodman, R.E., and J. Dubois, "Duplication of Dilatancy in Analysis of
       Jointed Rocks," Journal Soil Mechanics and Foundation Division,
       Proceedings ASCE, 1972, Vol. 98, No. SM 4, pp. 399-422.

[6]   Goodman, R.E., and St. John, "Static Finite Element Analysis of Jointed
       Rock," Numerical Methods in Geotechnical Engineering, McGraw
       Hill, Christian and Desai, Editors, in preparation.

[7]   Malina, H., "The Numerical Determination of Stresses and Deformations in
       Rock Taking Into Account Discontinuities," Proceedings 19th
       Colloquium on Geomechanics, Salzburg, 1969.

[8]   Zinkiewicz, O.C., The Finite Element Method in Structural and Contiuum
       Mechanics, McGraw Hill, 1967, p. 272.

[9]   Zinkiewicz, O.C., B. Best, C. Dullage, and K. Stagg, "Analysis of Non-Linear
       Problems in Rock Mechanics with Particular Reference to Jointed
       Rock Systems," Proceedings 2nd Congress ISRM, Belgrade, 1970,
       Vol. 3, Paper 8-14.

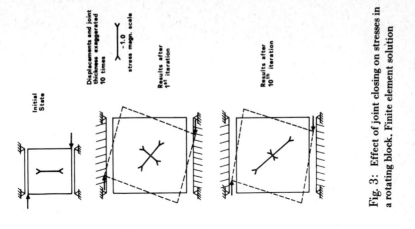

Fig. 3: Effect of joint closing on stresses in a rotating block. Finite element solution

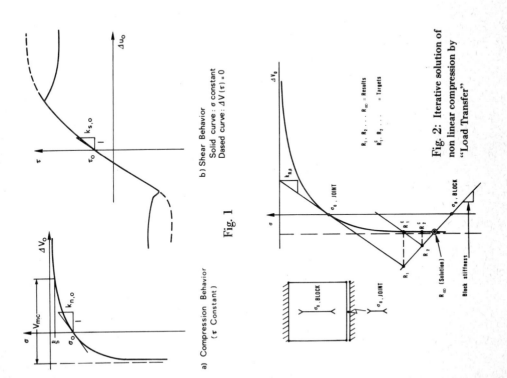

Fig. 1

a) Compression Behavior (τ Constant)

b) Shear Behavior
Solid curve : σ constant
Dased curve: $\Delta V(\tau) = 0$

Fig. 2: Iterative solution of non linear compression by "Load Transfer"

$R_1, R_2, \ldots R_\infty$ = Results
$R_1^c, R_2^c, \ldots$ = Targets

ROCK ANISOTROPY
Theory and Laboratory Testing

G. BARLA
Studio Geodetico Italiano

---

* All figures quoted in the text are at the end of the lecture.

Summary

Rock is frequently treated as a linearly elastic, homogeneous and isotropic medium. However, this particular behavior provides only limited insight into the true character of stresses and deformations within a rock mass.

A well-known behavior pattern which is of considerable interest in many engineering applications is the rock anisotropy. Significant errors can be introduced in stress and deformation analyses by assuming anisotropic rocks to be isotropic.

Many Authors have given careful attention to this problem. It is the purpose of the two lectures to describe some of the efforts made in recent years as confined to anisotropic rocks.

Introduction

In engineering analyses, rock is frequently treated as a linearly elastic, homogeneous, and isotropic medium. However, rock material exhibits, in most cases, different anomalies in the physical behavior, which cannot be neglected if a complete solution of any given rock mechanics problem is to be attained.

A well-known behavior pattern which is of considerable interest in many applications is the rock anisotropy. This behavior which involves deformation and strength characteristics, is not only confined to laminated or layered rocks. There is sufficient experimental evidence to support that even some granites can exhibit anisotropic behavior (Peres Rodrigues, 1966, Douglass and Voight, 1969).

It has been shown that anisotropy can be of relevance in many rock engineering applications. For example, significant errors are introduced in stress determinations by assuming anisotropic rocks to be isotropic (Berry and Fairhurst, 1966; Becker and Hooker, 1967; Barla and Wane, 1970). Further, when determining the deformation moduli in stratified rock masses by in situ arrangements, the use of analytical formulations based upon elastic anisotropic theory is worth of attention (Oberti et Al., 1970).

In recent years, many Authors have devoted a considerable effort to the problem of rock anisotropy, both from theoretical and experimental point of view. At first, this could appear a little out of the rock mechanics perspectives, as the continuum mechanics theory is generally called upon in order to produce the abstraction and formulate the solution to a given problem.

However, it is generally recognized that with the use of numerical techniques (e.g. finite element method), some of the restrictions met in the solution of rock engineering problems by continuum theory have now been overcome, so that the rock mass behavior can be described also including the influence of discontinuities. The necessity however remains to develop an appropriate model of behavior for the rock material. It is mainly with respect to such a model that the anisotropy is of relevance.

Since with continuum theories we introduce a model of behavior, the experiment is needed in order that this behavior can be substantiated or refuted. At least, we should be able to determine how much information is contained in our model which can be used for reliable design purposes.

The present lectures comprise some of the aspects which are of interest in the study of rock anisotropy. A very brief view of the theoretical concepts involved, confined only to deformation, will be given, as the aspects on strength will be discussed by other Lecturers. Some of the results on rock anisotropy, related to laboratory work, will be given. Finally, an attempt will be made to indicate specific areas in the field of rock anisotropy where further study is needed.

Theory
(a) Constitutive equations

Under the assumption of linear elasticity, a description of the rock material behavior is obtained with the following constitutive equation (generalized Hooke's law in cartesian coordinate x, y, z)

$$
\begin{Bmatrix} \epsilon_x \\ \epsilon_y \\ \epsilon_z \\ \gamma_{xy} \\ \gamma_{yz} \\ \gamma_{xz} \end{Bmatrix} =
\begin{bmatrix}
s_{11} & s_{12} & s_{13} & s_{14} & s_{15} & s_{16} \\
s_{21} & s_{22} & s_{23} & s_{24} & s_{25} & s_{26} \\
s_{31} & s_{32} & s_{33} & s_{34} & s_{35} & s_{36} \\
s_{41} & s_{42} & s_{43} & s_{44} & s_{45} & s_{46} \\
s_{51} & s_{52} & s_{53} & s_{54} & s_{55} & s_{56} \\
s_{61} & s_{62} & s_{63} & s_{64} & s_{65} & s_{66}
\end{bmatrix}
\begin{Bmatrix} \sigma_x \\ \sigma_y \\ \sigma_z \\ \tau_{xy} \\ \tau_{yz} \\ \tau_{xz} \end{Bmatrix}
\tag{1}
$$

where s_{ij} are the 36 elastic constants, ϵ_{ij} and σ_{ij} are the strain and stress tensors. The 36 constants s_{ij} are readily reduced to 21. However, by introducing laws of symmetry for the elastic properties, the number of independent constants is further reduced.

In the present discussion, two cases of elastic symmetry are of interest: a) the orthotropic material (i.e. 3 planes of elastic symmetry which result in 9 material constants), b) the transversely isotropic material (i.e. a plane of elastic symmetry, an axis of rotation, which result in 5 material constants).

For these materials, equation (1) let one write in terms of the engineering elastic constants (Young's moduli and Poisson's ratios) (Lekhnitskii, 1963):

(a1) Orthotropic material (Fig. 1)

$$(2) \quad \begin{Bmatrix} \epsilon_x \\ \epsilon_y \\ \epsilon_z \\ \gamma_{xy} \\ \gamma_{yz} \\ \gamma_{xz} \end{Bmatrix} = \begin{bmatrix} \dfrac{1}{E_1} & -\dfrac{\nu_{21}}{E_2} & -\dfrac{\nu_{31}}{E_3} & 0 & 0 & 0 \\[2mm] -\dfrac{\nu_{12}}{E_1} & \dfrac{1}{E_2} & -\dfrac{\nu_{32}}{E_3} & 0 & 0 & 0 \\[2mm] -\dfrac{\nu_{13}}{E_1} & -\dfrac{\nu_{23}}{E_2} & \dfrac{1}{E_3} & 0 & 0 & 0 \\[2mm] 0 & 0 & 0 & \dfrac{1}{G_{12}} & 0 & 0 \\[2mm] 0 & 0 & 0 & 0 & \dfrac{1}{G_{23}} & 0 \\[2mm] 0 & 0 & 0 & 0 & 0 & \dfrac{1}{G_{13}} \end{bmatrix} \begin{Bmatrix} \sigma_x \\ \sigma_y \\ \sigma_z \\ \tau_{xy} \\ \tau_{yz} \\ \tau_{xz} \end{Bmatrix}$$

where

E_i (E_1, E_2, E_3) = modulus of elasticity in i (x, y, z) direction.

ν_{ij} (ν_{12}, ν_{21}, ν_{13}, ν_{31}, ν_{23}, ν_{32}) = Poisson's ratio – ratio of strain in j direction to strain in i direction, due to stress in i direction.

G_{ij} (G_{12}, G_{13}, G_{23}) = shear modulus in plane i-j.

and

$$E_1 \; \nu_{21} \; = \; E_2 \; \nu_{12}$$

$$E_2 \quad \nu_{32} \quad = \quad E_3 \quad \nu_{23}$$
$$E_3 \quad \nu_{13} \quad = \quad E_1 \quad \nu_{31}$$

(3)

(a2) Transversely isotropic material (Fig. 2)

$$
\begin{Bmatrix} \epsilon_x \\ \epsilon_y \\ \epsilon_z \\ \gamma_{xy} \\ \gamma_{yz} \\ \gamma_{xz} \end{Bmatrix}
=
\begin{bmatrix}
\dfrac{1}{E_1} & -\dfrac{\nu_1}{E_1} & -\dfrac{\nu_2}{E_2} & 0 & 0 & 0 \\[2mm]
-\dfrac{\nu_1}{E_1} & \dfrac{1}{E_1} & -\dfrac{\nu_2}{E_2} & 0 & 0 & 0 \\[2mm]
-\dfrac{\nu_2}{E_2} & -\dfrac{\nu_2}{E_2} & \dfrac{1}{E_2} & 0 & 0 & 0 \\[2mm]
0 & 0 & 0 & \dfrac{1}{G_1} & 0 & 0 \\[2mm]
0 & 0 & 0 & 0 & \dfrac{1}{G_2} & 0 \\[2mm]
0 & 0 & 0 & 0 & 0 & \dfrac{1}{G_2}
\end{bmatrix}
\begin{Bmatrix} \sigma_x \\ \sigma_y \\ \sigma_z \\ \tau_{xy} \\ \tau_{yz} \\ \tau_{xz} \end{Bmatrix}
$$

(4)

where:

E_1 = modulus of elasticity in all directions lying in the plane of isotropy (xy plane);

E_2 = modulus of elasticity in direction z perpendicular to the plane of isotropy

ν_1 = Poisson's ratio in the plane of isotropy;

ν_2 = Poisson's ratio, i.e. ratio of strain in the plane of isotropy to strain in direction normal to it, due to stress in this same direction;

G_1 and G_2 = shear moduli in planes respectively parallel and normal to the plane of isotropy.

(a3) Stratified media (Fig. 3)

Many rocks are layered media. Under certain conditions their behavior can be represented by an "equivalent", homogeneous, and transversely isotropic medium (Pinto, 1966; Salamon, 1968; Wardle and Gerrard, 1972).

The following assumptions are made:

1) all layers are bounded by parallel planes and no relative displacement takes place on these planes;

2) all layers are homogeneous, transversely isotropic, with thickness and elastic properties which vary randomly with respect to the direction perpendicular to bounding planes;

3) a representative sample of the layered medium must contain a large number of layers.

It is shown that if all the layers are isotropic, the five independent elastic constants of the equivalent transversely isotropic medium are (Salamon, 1968):

$$\nu_1 = \frac{\sum \dfrac{\varphi_i \, \nu_{1i} \, E_{1i}}{1 - \nu_{1i}^2}}{\sum \dfrac{\varphi_i \, E_{1i}}{1 - \nu_{1i}^2}}$$

$$\nu_2 = (1 - \nu_1) \sum \frac{\varphi_i \, \nu_{2i}}{1 - \nu_{1i}}$$

$$E_1 = (1 - \nu_1^2) \sum \frac{\varphi_i \, E_{1i}}{1 - \nu_{1i}^2}$$

(5)

$$E_2 = \frac{1}{\sum \dfrac{\varphi_i}{E_{1i}} \left(\dfrac{E_{1i}}{E_{2i}} - \dfrac{2 \, \nu_{2i}^2}{1 - \nu_{1i}} \right) + \dfrac{2 \, \nu_2^2}{(1 - \nu_1) E_1}}$$

$$G_1 = \sum \varphi_i \, G_{1i}$$

$$G_2 = \frac{1}{\sum \dfrac{\varphi_i}{G_{2i}}}$$

(In all cases summation from $i = 1$, n is implied; n = number of layers) where:

E_{1i}, E_{2i} and ν_{1i}, ν_{2i} are the moduli of elasticity and the Poisson's ratios of the i-th layer;

$t_i = \varphi_i L$ is the thickness of the i-th layer;

L is the edge dimension of the cube;

φ_i,   as defined above implies that $\Sigma\varphi_i = 1$.

The following inequalities are shown to hold for the elastic constants of the equivalent medium (Wardle and Gerrard, 1972)

$$\begin{aligned}
E_1 &> 0 \\
E_2 &> 0 \\
\frac{G_1}{G_2} &\geqslant 1 \\
G_2 &> 0 \\
-1 \leqslant \nu_1 &\leqslant \frac{1}{2} \\
-1 \leqslant \nu_2 &< 2
\end{aligned} \tag{6}$$

As an example let us consider the case of fig. 4, where the "equivalent" medium is formed by two isotropic materials, say a (E_a, ν_a) and b (E_b, ν_b). Define A_a and A_b as the volume of each material in a unit volume of the equivalent, transversely isotropic medium. This obviously means that, for $A_a = 1$ and $A_b = 0$, the equivalent material goes into material a alone.

By using equations (5), the ratios E_1/E_a, E_2/E_a, and G_2/E_a have been evaluated for E_a/E_b = constant, $\nu_a = \nu_b = 0.25$, as A_a varies between 0.05 and 1. The numerical results are reported in figures 5, 6, 7. It is noticed that E_1/E_a is linearly related to A_a. On the contrary, a non linear relation to A_a occurs for E_2/E_a and G_2/E_a.

(a4) **Transformation of material constants**

In an anisotropic medium, the material constants depend on the direction of the coordinate axes of the orthogonal system used for reference. For the isotropic medium only, the material constants are invariant in any coordinate system.

In the theory of anisotropic media (Lekhnitskii, 1963), the elastic constants are shown to transform according with tensorial laws. In particular, consider a transversely isotropic medium, referred to the x, y, z system, with the z axis chosen as the axis of rotational symmetry (Fig. 8). With respect to the $\bar{x}, \bar{y}, \bar{z}$ system, rotated through the angle β about the x axis, equation (1) writes

$$
(7) \quad
\begin{Bmatrix}
\bar{\epsilon}_x \\
\bar{\epsilon}_y \\
\bar{\epsilon}_z \\
\bar{\gamma}_{xy} \\
\bar{\gamma}_{yz} \\
\bar{\gamma}_{xz}
\end{Bmatrix}
=
\begin{bmatrix}
\bar{s}_{11} & \bar{s}_{12} & \bar{s}_{13} & \bar{s}_{14} & \bar{s}_{15} & \bar{s}_{16} \\
\bar{s}_{12} & \bar{s}_{22} & \bar{s}_{23} & \bar{s}_{24} & \bar{s}_{25} & \bar{s}_{26} \\
\bar{s}_{13} & \bar{s}_{23} & \bar{s}_{33} & \bar{s}_{34} & \bar{s}_{35} & \bar{s}_{36} \\
\bar{s}_{14} & \bar{s}_{24} & \bar{s}_{34} & \bar{s}_{44} & \bar{s}_{45} & \bar{s}_{46} \\
\bar{s}_{15} & \bar{s}_{25} & \bar{s}_{35} & \bar{s}_{45} & \bar{s}_{55} & \bar{s}_{56} \\
\bar{s}_{16} & \bar{s}_{26} & \bar{s}_{36} & \bar{s}_{46} & \bar{s}_{56} & \bar{s}_{66}
\end{bmatrix}
\begin{Bmatrix}
\bar{\sigma}_x \\
\bar{\sigma}_y \\
\bar{\sigma}_z \\
\bar{\tau}_{xy} \\
\bar{\tau}_{yz} \\
\bar{\tau}_{xz}
\end{Bmatrix}
$$

where

$\bar{s}_{ij}, \bar{\epsilon}_{ij}$, and $\bar{\sigma}_{ij}$ are respectively the elastic constants, the strain and stress tensors, referred to the $\bar{x}, \bar{y}, \bar{z}$ system.

The $\bar{s}_{ij}$ are related to the eslatic constant, referred to the x, y, z system, according to the following formulae

$$\bar{s}_{11} = \frac{1}{E_1}$$

$$\bar{s}_{12} = - \left(\frac{\nu_1}{E_1} \cos^2 \beta + \frac{\nu_2}{E_2} \sin^2 \beta \right)$$

$$\bar{s}_{13} = - \left(\frac{\nu_1}{E_1} \sin^2 \beta + \frac{\nu_2}{E_2} \cos^2 \beta \right)$$

$$\bar{s}_{15} = 2 \left(\frac{\nu_2}{E_2} - \frac{\nu_1}{E_1} \right) \sin \beta \cos \beta$$

$$\bar{S}_{22} = \frac{1}{E_1} \cos^4 \beta + \left(\frac{1}{G_2} - \frac{2\nu_2}{E_2} \right) \sin^2 \beta \, \cos^2 \beta + \frac{1}{E_2} \sin^4 \beta$$

$$\bar{S}_{23} = \left(\frac{1}{E_1} + \frac{1}{E_2} + \frac{2\nu_2}{E_2} - \frac{1}{G_2} \right) \sin^2 \beta \cdot \cos^2 \beta \quad - \frac{\nu_2}{E_2}$$

$$\bar{S}_{25} = 2 \left(\frac{\cos^2 \beta}{E_1} - \frac{\sin^2 \beta}{E_2} \right) - \left(\frac{1}{G_2} - \frac{2\nu_2}{E_2} \right) (\cos^2 \beta - \sin^2 \beta) \, \sin \beta \, \cos \beta$$

$$\bar{S}_{33} = \frac{1}{E_1} \sin^4 \beta + \left(\frac{1}{G_2} - \frac{2\nu_2}{E_2} \right) \sin^2 \beta \, \cos^2 \beta + \frac{1}{E_2} \cos^4 \beta$$

$$\bar{S}_{35} = 2 \left(\frac{\sin^2 \beta}{E_1} - \frac{\cos^2 \beta}{E_2} \right) + \left(\frac{1}{G_2} - \frac{2\nu_2}{E_2} \right) (\cos^2 \beta - \sin^2 \beta) \, \sin \beta \, \cos \beta$$

(8)

$$\bar{S}_{44} = \frac{2(1 + \nu_1)}{E_1} \cos^2 \beta + \frac{1}{G_2} \sin^2 \beta$$

$$\bar{S}_{46} = \left(\frac{2(1 + \nu_1)}{E_1} - \frac{1}{G_2} \right) \sin \beta \, \cos \beta$$

$$\bar{S}_{55} = 4 \left(\frac{1}{E_1} + \frac{1}{E_2} + \frac{2\nu_2}{E_2} - \frac{1}{G_2} \right) \sin^2 \beta \, \cos^2 \beta + \frac{1}{G_2}$$

$$\bar{S}_{66} = \frac{2(1 + \nu_1)}{E_1} \sin^2 \beta + \frac{1}{G_2} \cos^2 \beta$$

$$\bar{S}_{14} = \bar{S}_{16} = \bar{S}_{24} = \bar{S}_{26} = \bar{S}_{34} = \bar{S}_{36} = \bar{S}_{45} = \bar{S}_{56} = 0$$

Defining the elastic constants with respect to the $\bar{x}$, $\bar{y}$, $\bar{z}$ coordinate system as

$$\bar{E}_1 = \frac{\bar{\sigma}_x}{\bar{\epsilon}_x}, \quad \bar{E}_2 = \frac{\bar{\sigma}_y}{\bar{\epsilon}_y}, \quad \bar{E}_3 = \frac{\bar{\sigma}_z}{\bar{\epsilon}_z},$$

$$\bar{\nu}_{12} = -\frac{\bar{\epsilon}_y}{\bar{\epsilon}_x}, \quad \bar{\nu}_{21} = -\frac{\bar{\epsilon}_x}{\bar{\epsilon}_y},$$

$$\bar{\nu}_{13} = -\frac{\bar{\epsilon}_z}{\bar{\epsilon}_x}, \quad \bar{\nu}_{31} = -\frac{\bar{\epsilon}_x}{\bar{\epsilon}_z},$$

(9)
$$\bar{\nu}_{23} = -\frac{\bar{\epsilon}_z}{\bar{\epsilon}_y}, \quad \bar{\nu}_{32} = -\frac{\bar{\epsilon}_y}{\bar{\epsilon}_z},$$

$$\bar{G}_{12} = \frac{\bar{\tau}_{xy}}{\bar{\gamma}_{xy}},$$

$$\bar{G}_{13} = \frac{\bar{\tau}_{xz}}{\bar{\gamma}_{xz}},$$

$$\bar{G}_{23} = \frac{\bar{\tau}_{yz}}{\bar{\gamma}_{yz}},$$

the following relations can be obtained

$$\bar{E}_1 = E_1$$

$$\bar{E}_2 = \frac{E_1}{\cos^4\beta + \frac{E_1}{E_2}\sin^4\beta + \left(\frac{E_1}{G_2} - 2\nu_2\frac{E_1}{E_2}\right)\sin^2\beta\,\cos^2\beta}$$

$$\bar{E}_3 = \frac{E_1}{\sin^4\beta + \frac{E_1}{E_2}\cos^4\beta + \left(\frac{E_1}{G_2} - 2\nu_2\frac{E_1}{E_2}\right)\sin^2\beta\,\cos^2\beta}$$

$$\bar{\nu}_{12} = E_1 \left(\frac{\nu_1}{E_1} \cos^2\beta + \frac{\nu_2}{E_2} \sin^2\beta \right)$$

$$\bar{\nu}_{23} = \frac{\bar{E}_1}{E_1} \left[\nu_2 \frac{E_1}{E_2} - \left(1 + \frac{E_1}{E_2} + 2\nu_2 \frac{E_1}{E_2} - \frac{E_1}{G_2} \right) \sin^2\beta \ \cos^2\beta \right] \qquad (10)$$

$$\bar{\nu}_{13} = E_1 \left(\frac{\nu_1}{E_1} \sin^2\beta + \frac{\nu_2}{E_2} \cos^2\beta \right)$$

$$\frac{G_2}{\bar{G}_{23}} = \frac{G_2}{E_1} \left[4 \left(1 + \frac{E_1}{E_2} + 2\nu_2 \frac{E_1}{E_2} - \frac{E_1}{G_2} \right) \sin^2\beta \ \cos^2\beta + \frac{E_1}{G_2} \right]$$

$$\frac{G_2}{\bar{G}_{12}} = \left[\sin^2\beta + 2 \ (1 + \nu_1) \frac{G_2}{E_1} \cos^2\beta \right]$$

$$\frac{G_2}{\bar{G}_{13}} = \left[\cos^2\beta + 2 \ (1 + \nu_1) \frac{G_2}{E_1} \sin^2\beta \right]$$

and

$$\bar{\nu}_{12} \ \bar{E}_2 = \bar{\nu}_{21} \ \bar{E}_1$$

$$\bar{\nu}_{23} \ \bar{E}_3 = \bar{\nu}_{32} \ \bar{E}_2 \qquad (11)$$

$$\bar{\nu}_{13} \ \bar{E}_3 = \bar{\nu}_{31} \ \bar{E}_1$$

As simple examples of application of the above equations (10), we can consider the following cases:

a) $\qquad \frac{E_1}{E_2} = 2 \ , \ \nu_2 = 0$

b) $\qquad \frac{E_1}{E_2} = 2 \ , \ \nu_2 = 0.25$

The Fig. 9 and 10 illustrate the change of $E_1/\bar{E}_2$ with β, for $E_1/G_2 =$ = constant. It is of interest to observe that, for $\nu_2 \neq 0$, values of $\bar{E}_2$ greater than E_1 may even occur for certain angles β.

Laboratory testing

Laboratory testing of anisotropic rocks has been carried out in recent years by considering:

1) Rocks which would be expected to behave as isotropic media and however exhibit anisotropic properties (class A);
2) Rocks which are clearly anisotropic in nature and show directions of symmetry for their strength and deformation characteristics (class B).

Two different approaches are generally applied. With the first one, the principal axes of strength and deformation symmetry must be found; thus a representative surface is defined which allows one to describe the spatial variation of the material constants with orientation. With the second one, the axes of symmetry are assumed as known a priori and the material constants along these same axes are determined.

The following testing methods will be briefly discussed:

(a) uniaxial compression test
(b) uniaxial tension test
(c) indirect tensile test.

Mainly, experimental results are referred to rather than description of technical details on the equipment used.

(a) Uniaxial compression test

The uniaxial compression test is the simplest and most widely used method for determining the strength and deformation characteristics of isotropic rock. Thus, the same method has been preferentially applied to the study of rock anisotropy.

(a1) Class A rocks

When consideration is given to the class A rocks the following approach is generally used (Peres Rodrigues, 1966; Douglass and Voight, 1969).

A square block is cut from a given site and either cylindrical or prismatic specimens are sampled in different orientations (Fig. 11) with respect to reference axes. For each specimen uniaxial compression tests are performed by determining the tangent (E_t) and secant (E_s) moduli, and the compressive strength (C_o) (*).

(*) Microscopic fabric observations on thin sections in the laboratory and structural investigations in situ are the ./.

In this investigation the moduli and compressive strength are assumed to vary with orientation according to the following ellipsoid-type quadratic law (*)

$$\{x \quad y \quad z\} \begin{bmatrix} a_1 & a_2 & a_3 \\ a_2 & a_4 & a_5 \\ a_3 & a_5 & a_6 \end{bmatrix} \begin{Bmatrix} x \\ y \\ z \end{Bmatrix} = 1 \qquad (12)$$

where x, y, and z are vector components in the reference coordinate direction and a_i, $i = 1$ to 6, are coefficients of the quadric.

These coefficients can be determined by the method of least squares, given the experimental values of the material constants in different directions. Then, the characteristic roots are found together with the directional cosines $\alpha_i, \beta_i, \gamma_i$ of the ellipsoid principal axes $\bar{x}, \bar{y}, \bar{z}$ with respect to the x, y, z reference coordinate system (Fig. 11).

The following equations can therefore be written for the ellipsoids in the standard form:

for E_t and E_s,

$$\frac{\bar{x}^2}{A^2} + \frac{\bar{y}^2}{B^2} + \frac{\bar{z}^2}{C^2} = 1 \qquad (13)$$

for C_o,

$$\frac{\bar{x}_1^2}{A_1} + \frac{\bar{y}_1^2}{B_1} + \frac{\bar{z}_1^2}{C_1} = 1 \qquad (14)$$

where A, B, C and A_1, B_1, C_1 are the semi-axes; $\bar{x}, \bar{y}, \bar{z}$ and $\bar{x}_1$, $\bar{y}_1$, and $\bar{z}_1$ are the principal axes respectively for the ellipsoid of deformation moduli and strength.

A comparison of the "theoretical" (ellipsoid) values with experimental values provides a means for evaluating the initial assumption of an ellipsoid quadratic law. Additionally, a rotation matrix [R] which carries the corresponding axes of the two ellipsoids to coincidence can be defined. Also a relationship between the corresponding semi-axes of the ellipsoids allows one to find either a homotetic

---

./. associated studies carried out.

(*) The "direction" surface for Young's modulus is shown to be of the fourth order for the general case. When additional relations are assumed to hold among the elastic constants, the surface is an ellipsoid (Lekhnitskii, 1963).

ellipsoid (i.e. the ratios of the semi-axes are equal) or an autometric ellipsoid (i.e. otherwise).

By using the approach briefly summarized above, Peres Rodrigues (1966) and Douglass and Voight (1969) investigated the anisotropy of granites and the results they obtained are very promising.

Peres Rodrigues was able to draw the following main conclusions from his investigation:

1) The deformation moduli E_t and E_s and the compressive strength vary with orientation by following in granites very satisfactorily an ellipsoid-type quadratic law (Fig. 12).

2) The ellipsoids of deformation moduli and strength can be correlated by linear transformations composed of a comparatively small amplitude rotation and an autometric transformation.

3) The ratio of the major and minor semi-axis of the ellipsoid (defined as the maximum anisotropy ratio) gives average values of 1.75 for deformation moduli and 1.35 for compressive strength. Thus, the anisotropy of deformability seems to be more prevalent than the anisotropy of strength.

4) The axes of the ellipsoids are either parallel or normal to the attitudes of joint systems and slopes at the site where the blocks are taken from.

Douglass and Voight confirmed the results reported above as far as the validity of an ellipsoidal distribution for deformation moduli and strength. They made the following additional observations, based also on a comprehensive analysis which invloved microscopic fabric studies:

1) The anisotropy of granites can be considered to be an effect produced by microfractures and narrow cavities between adjacent particle boundaries in the rock.

2) The stress-strain curve for the same rocks, at low to moderate stress levels, is non linear and shows a concave upward curvature. Thus, non linear and anisotropic behaviors have the same origin.

3) The anisotropy of deformability is seen to decrease with increasing stress (Fig. 13, Table 1).

4) The directions of major moduli axes are related to quartz optic axes maxima; the directions of minimum compressive strength appear to be clearly related to preferred orientations of microfractures.

(a2) Class B rocks

When class B rocks are tested in uniaxial compression, the usual assumption is to accept the transversely isotropic behavior, which makes the plane of isotropy to coincide with the plane of schistosity and the axis of rotational symmetry to be the normal to this same plane (Dayre, 1969; Masure, 1970, Monhoie, 1970; Pinto, 1970) (*).

Either cylindrical or prismatic specimens are tested for uniaxial compression strength $(C_o)_\beta$, by varying the direction of application of the uniaxial load (i.e. angle β in Fig. 14). The results are generally reported in the form of Fig. 15, which refers to a serpentinous schist from Val Malenco (Sondrio, Italy).

Cylindrical specimens (diameter 29.7 mm, height = 2.5 diameter) were obtained from squared blocks. Different inclinations (angle β of Fig. 14) of the plane of schistosity with respect to the direction of the applied stress σ_z were considered by using three specimens for each orientation, at 15 degrees increment. The results of Fig. 15 reported for each value of β are average values.

One can notice that the minimum $(C_o)_\beta$ is obtained for $\beta = 60°$. Furthermore, the specimen generally fails along schistosity planes when $30 \leqslant \beta \leqslant 60°$ and $\beta = 90°$. For $0 \leqslant \beta \leqslant 15°$ and $\beta = 75°$ this is not observed and the rock matrix is mainly intersected by the plane of failure (see Fig. 16 where the specimens for $\beta = 0°$ and $\beta = 90°$ are shown as they appear just after testing).

When the rock is tested for deformation behavior, the main purpose is the determination of the 5 independent elastic constants E_1, E_2, ν_1, ν_2, and G_2, which define the transversely isotropic behavior.

If the uniaxial compression test is performed as shown in Fig. 17 A, by loading the specimen perpendicularly to the plane of isotropy, E_2 and ν_2 can be found to be

$$E_2 = \frac{\sigma_z}{\epsilon_z}$$
$$\nu_2 = -\frac{\epsilon_x}{\epsilon_z}$$

(15)

Conversely, by loading the specimen parallely to the plane of isotropy

---

(*) The work of Peres Rodrigues (1970) should be mentioned with respect to class B rocks. He assumes ortotropic behavior with unique axis known and applies again the theory of direction surfaces of higher degree than the second.

(Fig. 17 B), E_1 and ν_1 are given by

$$E_1 = \frac{\sigma_x}{\epsilon_x}$$

$$\nu_1 = -\frac{\epsilon_y}{\epsilon_x}$$

(16)

Thus, the measurement of longitudinal and transversal strains in the specimen allows the Young's moduli and Poisson's ratio to be determined with simplicity.

The fifth independent elastic constant G_2 (*) can be evaluated by testing the rock specimen as shown in Fig. 16, where $\beta = 45°$.

We have in this case

$$\tau_{yz} = \frac{P}{A} \sin\beta \cos\beta = \frac{1}{2}\frac{P}{A}$$

$$G_2 = \frac{\tau_{yz}}{\gamma_{yz}}$$

(17)

where γ_{yz} is the shear strain which can be determined with a 45 degree strain rosette in the yz plane.

Alternately, one can use the second of (10) to write

$$\frac{1}{G_2} = \frac{\dfrac{1}{\overline{E}_2} - \dfrac{\sin^4\beta}{E_1} - \dfrac{\cos^4\beta}{E_2}}{\sin^2\beta \, \cos^2\beta} + 2\,\nu_2\,\frac{1}{E_2}$$

(18)

which shows that G_2 can be determined by knowing E_1, E_2, and ν_2, if $\overline{E}_2$ is measured (i.e. the strain in the direction of loading is known). Obviously, if this second approach is used, the tensorial law for transformation of the material constant is accepted.

(*) The determination of G_2 is of extreme interest. Quite often, in practical applications one avoids to measure it. Simplifying assumptions are therefore introduced. It can however be shown that the usual condition

$$\frac{1}{G_{ij}} = \frac{1 + \nu_{ij}}{E_i} + \frac{1 + \nu_{ji}}{E_j}$$

(which reduces respectively to six and four the independent elastic constants for the orthotropic and transversely isotropic material) is not acceptable for many rocks (Martino and Ribacchi, 1972).

An example of determination of the elastic constants can be reported by referring to some experimental results given by Pinto (1970). Fig. 18 shows the variation of the elastic constant $\bar{E}_2$ with β. The experimental results are compared with the theoretical prediction based upon tensorial laws of transformation.

(b) Uniaxial tensile test

The strength and deformation properties of rock when subjected to tensile stress are difficult to determine.

This fact justifies the limited number of experimental results presently available on the subject, even when isotropic rocks are considered. Indirect methods are often used in practice, but their application to anisotropic rocks poses some difficulties in providing appropriate formulae for interpretation of results (*). Furthermore, if consideration is to be given to the deformation behavior, the use of the uniaxial test seems to be the most appropriate one.

The axial application of the tensile load and the transmission of the tensile stress, without inducing anomalous stress concentrations in the specimen, are two aims, difficult to achieve in the laboratory.

The use of an elastic joint, similar to that proposed by Dubois (1970), and the application of a carefully shaped cap, to which the specimen is cemented with epoxy resin (Fig. 19), allow the tensile test to be performed correctly in the laboratory (Barla and Goffi, 1973). Thus, the preliminary tests described below could be carried out on a gneiss from Entracque (Val Gesso, Italy), Fig. 20.

Cylindrical specimens (diameter 29.7 mm, height = 2.5 diameter) were drilled from squared blocks so that different inclinations (angle β in Fig. 14) of the plane of laminations could be considered. Three specimens were obtained for each value of β, at 15 degrees increment, except for β = 15 and 75 degrees.

The longitudinal (ϵ_1) and transversal (ϵ_3) strains are measured during testing up to failure, by means of electrical strain gauges.

Besides the limited number of tests carried out up to the present, the tangent modulus $\bar{E}_2$ in the direction of loading has been evaluated. The results are reported in Fig. 21, where a comparison is made with the theoretical prediction based upon the second equation of (10) and given values of E_1, E_2, E_1/G_2, and ν_2.

Additionally, it should be noted that in all tests carried out the

---

(*) An attempt in using the indirect methods for determining the tensile strength of anisotropic rocks is discussed below.

transversal strain exhibits the same sign of the corresponding longitudinal strain. This fact, which is as well observed in uniaxial compression tests, deserves attention and should be investigated thoroughly.

The tensile strength $(T_o)_\beta$ determined for the same rock, as a function of the angle β is reported in Fig. 22. The ratio $(T_o)_{90}/(T_o)_\beta$, where $(T_o)_{90}$ is the tensile strength for the specimen loaded parallel to the laminations [$(T_o)_{90} = 73$ kg/cm^2], is given. A photograph of two specimens, as they appear after failure, is shown in Fig. 23.

(c) Indirect tensile testing

The diametral compression of circular discs and tings (Fig. 23) is a well known method for determining the tensile strength of rock materials. This method is generally applied to homogeneous and isotropic rocks. The formulae used for calculation are derived for isotropic elasticity only.

Experimental evidence on the behavior of anisotropic rocks, when subjected to indirect tensile stress is limited (Hobbs, 1964). Further, only a few theoretical investigations have been reported on the subject (Fine and Vouille, 1970; Barron, 1971).

We have recently carried out a series of indirect tensile tests on two rock types which under uniaxial compression exhibit anisotropy both of deformability and strength (Barla and Innaurato, 1972): a granitoidic gneiss from Valle di Susa (Susa, Italy) and a serpentinous schist from Val Malenco (Sondrio, Italy). It is of interest to discuss the results obtained also in the light of some theoretical considerations.

(c 1) Diametral compression of circular discs

A total of 52 discs (24 of granitoidic gneiss and 28 of serpentinous schist), 56 mm diameter (D) and 11 mm thickness (t), were tested in a suitable system which insured the diametral load P be applied carefully and uniformly. The specimens were loaded at different orientations, with respect to the laminations, defined in the following with the angle β of Fig. 23.

The results for these tests are illustrated in Fig. 24 for the gneiss and in Fig. 25 for the schist. The ratio P_f/Dt (P_f is the value of P applied at failure) is shown to depend in both cases upon the value of β. It is of interest to notice that the maximum value of P_f/Dt occurs for $\beta = 0°$. The minimum value is found at $\beta = 60°$ for the schist and at $\beta = 75°$ for the gneiss.

The two rocks exhibit under testing different trends of behavior. The gneiss fails mostly along the diameter of load application (Fig. 26). On the contrary, the schist is seen often to fail along the laminations (Fig. 26). This observation raises doubts on the nature of the failure process.

This first series of experiments let us think that the tensile testing of anisotropic rocks by diametral loading of circular discs might be a suitable technique only for rocks which exhibit low anisotropy (e.g. the granitoidic gneiss).

However, even in this case a theoretical relationship must be established in order to define the dependence of the tensile strength upon β. In fact, the known formula for tensile strength calculation ($T_o = 2P_f / \pi Dt$), based upon isotropic elasticity, is no longer applicable.

(c2) Diametral compression of circular rings

The result obtained for the schist asked for a differential method of testing to be used in order to induce a tensile failure in this rock.

The diametral compression of circular rings was employed. 51 circular rings with 56 mm outside diameter (D_1), 34 mm inside diameter (D_2), and 11 mm thickness (t), were used. With the main purpose of ascertaining the influence of thickness, circular rings with the same cross section dimensions but 24 mm thickness were also tested. The results are illustrated in Fig. 27, where the values of $P_f / D_1 t$ are reported versus the angle β.

It is noticed that the ring thickness influences the results only slightly, as a maximum 10 per cent deviation in the value for $P_f / D_1 t$ is found among the two series of tests. A minimum for the curve $P_f / D_1 t$ vs. β occurs at $\beta = 75°$, when t = 11 mm. The rings with a 24 mm thickness show, however, a minimum at $\beta = 90°$.

The rings with smaller thickness is observed generally to fail first along the line of loading. Subsequently, new fractures arise along laminations and, in some cases, fractures develop at 5-15 degrees with respect to the laminations (Fig. 28).

It can tentatively be concluded that the rocks which are markedly anisotropic (e.g. the serpentinous schist) can be tested for tensile strength determination by using the diametral compression of circular rings.

(c3) Theoretical considerations

By using the finite element method, the tensile stress at the disc center was evaluated, under the assumptions of linear elasticity and by considering the rock as characterized by $E_1/E_2 = 1.12$, $\nu_1 = \nu_2 = 0.21$, for different values of β. The results are reported in Table 2, where the corresponding tensile stress at failure is also shown.

By the same method, and considering $E_1/E_2 = 2.0$, $\nu_1 = 0.34$, $\nu_2 = 0.12$ the tensile stress at the inner surface of the ring as a function of β was calculated. The results are reported in Table 3, where in each case the tensile stress at failure is given.

(c4) Discussion

The assumption was made that the anisotropy of deformability is a relevant factor in determining the distribution of stress in the rock specimen. The finite element method was used as shown above, in order to determine the tensile strength for anisotropic rocks as function of the angle β .

One can notice that if the procedure used above is accepted high values of T_o are obtained when the ring test is used. This fact certainly raises doubts on the testing method if T_o is to be considered a true material property.

Conclusions

We have briefly examined, both from theoretical and experimental points of view, the problem of rock anisotropy. No doubts, the results presently available on the subject are limited and it is difficult to draw any general conclusion from them. Further, our considerations were mainly concerned with the uniaxial test and the indirect tensile tests, for which results were available.

At the present stage the following areas, in the field of rock anisotropy, where further studies are recommended, can be indicated.

1) The extent to which the constitutive equations for a linearly elastic, homogeneous and transversely isotropic medium are applicable to anisotropic rocks is to be investigated further. Also the validity of tensorial laws in the description of the variation of material constants with respect to known directions of elastic symmetry is a point of relevant interest.

2) The anisotropic nature of rocks, which would be expected to behave apparently as isotropic materials, is to be explained in the light of extended studies on rock fabrics and structure. The relevance of it on the methods of determining the deformation moduli and the ground stress tensor is for example a related field of endeavor.

3) Rocks which are markedly anisotropic (e.g. layered, schistous, and stratified

rocks) should be carefully studied in order to see if the main directions of symmetry which can a priori be established for them can be correctly assumed to be also directions of symmetry for the material constants (i.e. the plane of schistosity and the direction normal to it can be assumed respectively to be plane of isotropy and axis of rotational symmetry).

4) The nonlinearly elastic behavior which many anisotropic rocks are shown to exhibit even at low stresses deserves interest, in order to see if it is a directional property, as it would seem to be on the basis of the experimental results presently available.

5) Other testing methods, mainly triaxial and poliaxial compression tests, should be investigated. Deformation and strength behaviors are to be determined with consideration being given to: a) the conditions of loading; b) the direction of the applied stresses with respect to relevant structural symmetries.

6) When using the numerical methods for solution of rock engineering problems, the constitutive equations for anisotropic media, as given in continuum mechanics, should not be regarded as the only means for the description of rock behavior. Other equations to be used for the characterization of rocks, including even the influence of anisotropy and non linearity, should be proposed.

REFERENCES

[1]   Barla, G. and Goffi, L. Prove di trazione diretta su roccia, Secondo
        Congresso Nazionale AIAS, Genova, 1973.

[2]   Barla, G. and Innaurato, N., Indirect tensile testing of anisotropic rocks,
        Rock Mechanics, 1972 (to be published).

[3]   Barla, G. and Wane, M.T., Stress-relief method in anisotropic rocks by means
        of gauges applied to the end of a borehole, Int. J. Rock Mech. Min.
        Sci., Vol. 7, pp. 171-182, 1970.

[4]   Barron, K., Brittle fracture initiation in and ultimate failure of rocks, Part III
        — Anisotropic rocks: experimental results, Int. J. Rock Mech. Min.
        Sci., Vol. 8, pp. 565-575, 1971.

[5]   Becker, R.M. and Hooker, V.E., Some anisotropic considerations in rock
        stress determinations, U.S.B.M., Report of Investigations 6965,
        1967.

[6]   Berry, D.S. and Fairhurts, C., Influence of rock anisotropy and time
        dependent deformation on the stress-relief and high-modulus in-
        clusion techniques of in situ stress determination, Testing Technique
        for Rock Mechanics, ASTM STP 402, p. 190, 1966.

[7]   Dayre, M., Discontinuous anisotropy of a limestone formation, Colloquium
        on Fissuring in Rocks, Rev. Ind. Minerale, Spec. Issue, pp. 35-40,
        1969.

[8]   Douglass, P.M. and Voight, B., Anisotropy of granites: a reflection of
        microscopic fabric, Geotechnique, 19, pp. 376-398, 1969.

[9]   Dubois, M., Experimental study of strain gauge high precision dynamometers
        at the O.N.E.R.A. Modane test centre, 1970.

[10] Fine, J. and Vouille, G., L'anisotropie des roches — Son influence sur l'essai brésilien, Revue de l'Industrie Minerale, pp. 5-12, 1970.

[11] Hobbs, D.W., The tensile strength of rocks, Int. J. Rock Mech. Min. Sci., Vol. 1, pp. 385-396, 1964.

[12] Lekhnitskii, S.G., Theory of elasticity of an anisotropic elastic body, Holden-Day, Inc., San Francisco, 1963.

[13] Martino, D. and Ribacchi, R., Osservazioni su alcuni metodi di misura delle caratteristiche elastiche di rocce o ammassi rocciosi, con particolare riferimento al problema dell'anisotropia, l'Industria Mineraria, pp. 193-203, 1972.

[14] Masure, P., Comportement mécanique des roches à anisotropie planaire discontinue, Proc. Second Congress of the International Society for Rock Mechanics, Beograd, 1970.

[15] Monjoie, A., Mechanical properties of the Silurian schists in Tihange, Belgium, Proc. Second Congress of the International Society for Rock Mechanics, Beograd, 1970.

[16] Oberti, G., Rebaudi, A. and Goffi, L., Comportement statistique des massifs rocheux (calcaires) dans la realisation de grands ouvrages sout-terrains, Proc. Second. Congress of the International Society for Rock Mechanics, Beograd, 1970.

[17] Peres Rodrigues, F., Anisotropy of granites. Modulus of elasticity and ultimate strength ellipsoids, joint systems, slope attitudes, and their correlations, Proc. First Congress of the International Society for Rock Mechanics, Lisbon, 1966.

[18] Peres Rodrigues, F., Anisotropy of rocks. Most probable surfaces of the ultimate stresses and of the moduli of elasticity, Proc. Second Congress of the International Society for Rock Mechanics, Beograd, 1970.

[19]   Pinto, L.J., Stress and strain in an anisotropic-orthotropic body, Proc. First
       Congress of the International Society for Rock Mechanics, Lisbon,
       1966.

[20]   Pinto, L.J., Deformability of schistous rocks, Proc. Second Congress of the
       International Society for Rock Mechanics, Beograd, 1970.

[21]   Salamon, M.D.G., Elastic moduli of a stratified rock mass. Int. J. Rock
       Mech. Min. Sci., Vol. 5, pp. 519-527, 1968.

[22]   Wardle, L.J. and Gerrard, C.M., The "equivalent" anisotropic properties of
       layered rock and soil masses, Int. J. Rock Mech. Min. Sci., Vol. 4, pp.
       155-175, 1972.

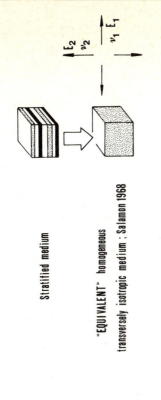

Figure 3 Stratified medium represented as "equivalent" homogeneous transversely isotropic material.

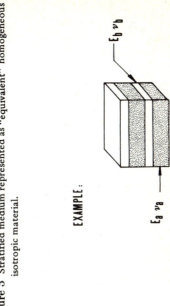

Figure 4 Stratified medium composed of two different materials with elastic constants E_a, ν_a, E_b, ν_b.

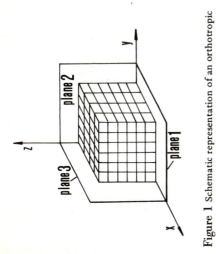

Figure 1 Schematic representation of an orthotropic material.

Figure 2 Schematic representation of a transversely isotropic material.

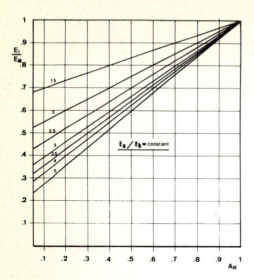

Figure 5 Variation of E_1/E_a, for given E_a/E_b and $\nu_a = \nu_b = 0.25$, with A_a. Two-layer medium represented as "equivalent" homogeneous transversely isotropic material.

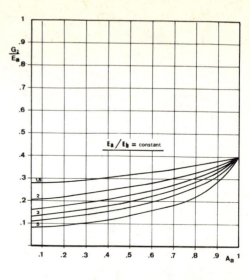

Figure 7 Variation of G_2/E_a, for given E_a/E_b and $\nu_a = \nu_b = 0.25$, with A_a. Two-layer medium represented as "equivalent" homogeneous transversely isotropic material.

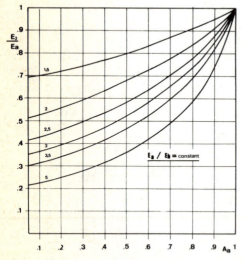

Figure 6 Variation of E_2/E_a, for given E_a/E_b and $\nu_a = \nu_b = 0.25$, with A_a. Two-layer medium represented as "equivalent" homogeneous transversely isotropic material.

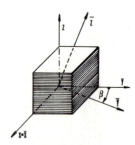

Figure 8 Coordinate systems used for studying the transformation law for material constants of a transversely isotropic medium.

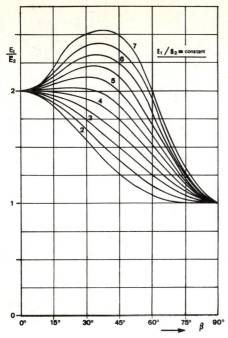

Figure 9 Variation of E_1/E_2 with β (for $E_1/G_2 =$ = constant, $\nu_2 = 0$) in a transversely iso-tropic material.

Figure 10 Variation of E_1/E_2 with β (for $E_1/G_2 =$ = 0.25) in a transversely isotropic material.

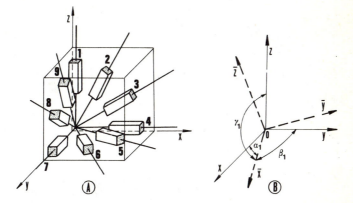

Figure 11 (A) Orientations of sampling in a block for determining the direction surfaces for strength and de-formability constants. (B) Reference coordinate axes (x, y, z) and ellipsoid principal axes ($\underline{x}$, $\underline{y}$, $\underline{z}$).

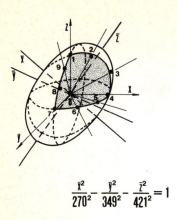

$$\frac{\bar{x}^2}{270^2} - \frac{\bar{y}^2}{349^2} - \frac{\bar{l}^2}{421^2} = 1$$

PERES RODRIGUES, 1966

Figure 12 Example of determination of the ellipsoid type quadratic law for a granite.

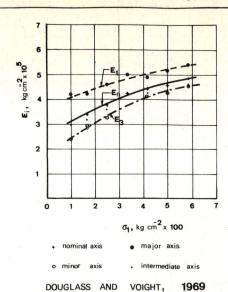

· nominal axis   • major axis

o minor axis   · intermediate axis

DOUGLASS AND VOIGHT, **1969**

Figure 13 Variation of major (E_1), intermediate (E_2) and
minor (E_3) ellipsoid principal semi-axes for tangent moduli.
Dashed lines represent approximate trends of E_1, E_3 and E_n,
where E_n, is the "nominal radius" of a volumetrically equivalent sphere.

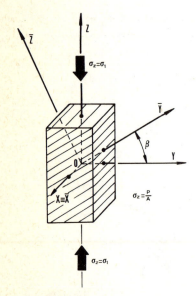

Figure 14 Schematic representation of the
uniaxial compression test for class B rocks
when the load is applied along z.

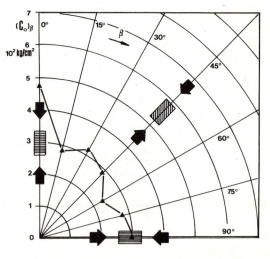

Figure 15 Variation of the uniaxial compressive strength $(C_o)_\beta$
vs. β for a serpentinous schist from Val Malenco (Sondrio, Italy).

A

B

Figure 16 Specimens of serpentinous schist after failure in uni-axial compression test. (A) $\beta = 0°$, (B) $\beta = 90°$.

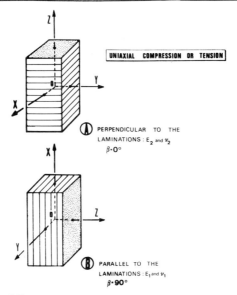

Figure 17 Schematic representation of the uniaxial compression
test for class B rocks when the load is applied: (A) perpendicular to
and (B) parallel to the plane of schistosity.

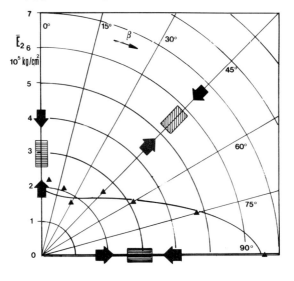

DATA DUE TO PINTO 1970

Figure 18 Variation of the elastic constant E_2 with β. Comparison
of experimental results and theoretical predictions (data obtained
for a schist by Pinto, 1970). Uniaxial compression test.

A

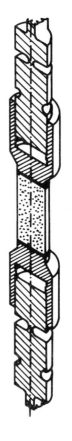

B

Figure 19 A view of the testing unit used in the uniaxial tensile test.

A

B

Figure 20 Specimens of a gneiss after failure in uniaxial tensile test.
(A) $\beta = 0°$, (B) $\beta = 90°$.

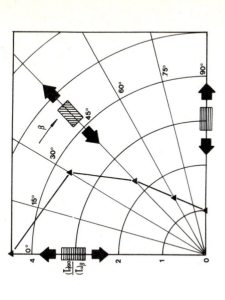

Figure 22 Variation of the tensile strength $(T_o)_\beta$, for a gneiss from Entracque (Val Gesso, Italy), vs. β. The ratio $(T_o)_{90}/(T_o)_\beta$ is shown.

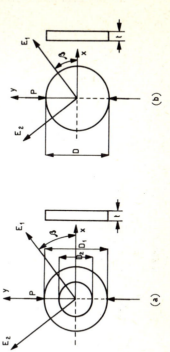

Figure 23 Diametral compression of: (A) circular ring, (B) circular disc.

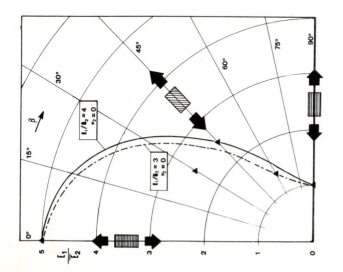

Figure 21 Variation of E_1/E_2 with β for a gneiss from Entracque (Val Gesso, Italy). Comparison of available experimental results and theoretical predictions. Uniaxial tensile test.

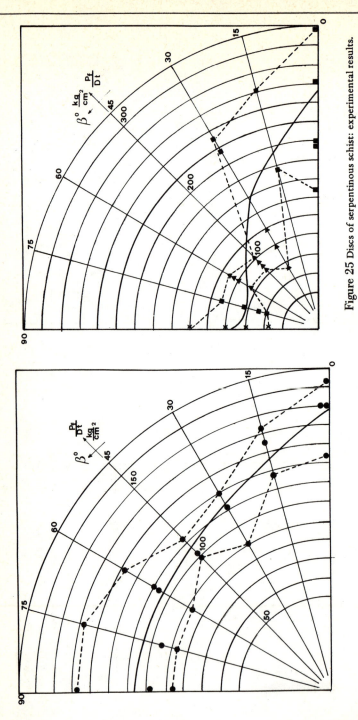

Figure 25 Discs of serpentinous schist: experimental results.
Polar diagrams, P_f/Dt vs. β.
X tensile failure ▲ shear failure ■ undefined failure

Figure 24 Discs of granitoidic gneiss: experimental results. Polar diagram,
P_f/Dt vs. β.

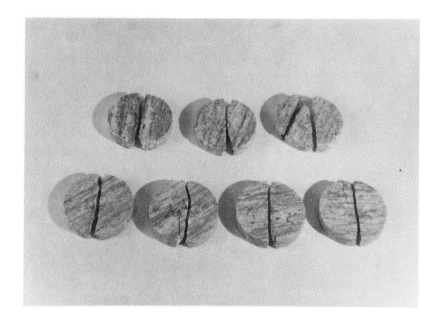

A

B

Figure 26 Circular discs after failure in diametral compression tests (indirect tensile tests).
(A) granitoidic gneiss from Entracque (Val Gesso, Italy),   (B)   serpentinous
schist from Val Malenco (Sondrio, Italy).

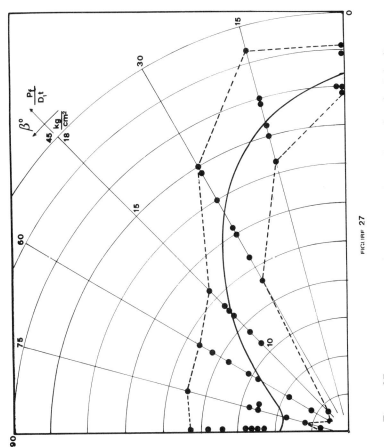

FIGURE 27

Figure 27 Rings of serpentinous schist (t = 11 mm): experimental results. Polar diagram,
$P_f/D_1 t$ vs. β.

A

B

Figure 28 Circular rings after failure in diametral compression tests (indirect tensile tests). Serpentinous schist from Val Malenco (Sondrio, Italy).

TABLE 1

Summary of anisotropy ellipsoid data for Barre granite

Stress level notation	Axial Stress bars	Young's moduli—tangent (bars x 10^5)				Anisotropy	
		$E_{T(1)}$	$E_{T(2)}$	$E_{T(3)}$	E_n	R_a	E_1/E_3
1	67	4.55	2.88	2.03	2.98	0.85	2.24
2	100	4.18	2.98	2.39	3.10	0.58	1.75
3	167	4.20	3.19	2.92	3.40	0.38	1.44
4	250	4.57	3.56	3.25	3.75	0.35	1.41
5	333	4.95	4.05	3.76	4.22	0.28	1.32
6	417	4.89	4.24	4.12	4.40	0.18	1.19
7	500	5.14	4.38	4.26	4.57	0.19	1.21
8	584	5.37	4.61	4.55	4.83	0.17	1.18
		Young's moduli—secant (bars x 10^5)					
		$E_{s(1)}$	$E_{s(1)}$	$E_{s(1)}$	E_n		
L	0–670	4.68	3.66	3.37	3.86	0.34	1.39
U*	0–670	4.90	4.05	3.62	4.15	0.31	1.35
		Compressive strength (bars)					
		C_1	C_2	C_3	C_n	R_a	C_1/C_3
C	—	2289	2001	1913	2080	0.18	1.20

$E_a = (E_1 E_2 E_3)^{1/3}$ $R_a = (E_1 - E_3)E_n$ * Unloading

TABLE 2

Finite element results for the disc ($E_1/E_2 = 1.12$)

$\beta(\cdot)$	Load at failure (P_f, kg)	Calculated tensile stress (σ_t, kg/cm^2) (*)	Calculated tensile stress at failure (T_0, kg/cm^2)	Parameter K_d
0 (isotropy)	890	127	93.0	0.634
15	760	121	74.0	0.605
30	690	117	75.0	0.590
45	655	113	59.0	0.565
60	635	110	58.5	0.550
75	640	109	55.8	0.590
90	650	107	64.0	0.590

(*) For the finite element computations the following is assumed: $P = 1000$ kg, $D = 50$ mm, $t = 10$ mm

TABLE 3

Finite element results for the ring ($E_2/E_2 = 2.0$)

$\beta(^\circ)$	Load at failure (P_f, kg)	Calculated tensile stress (σ_t kg/cm^2)(*)	Calculated tensile stress at failure (T_0, kg/cm^2)	Parameter K_r
0 (isotropy)	100	334	265	26.0
15	94	470	365	37.0
30	82	422	255	30,5
45	70	360	199	28.0
60	59	305	143	24.0
75	54	282	124	22.5
90	60	270	129	21.2

(*) For the finite element computations the following is assumed: $P = 100$ kg, $D_1/D_2 = 0,6$, $t = = 10$ mm.

ENGINEERING PROPERTIES
OF JOINTED ROCK

K. W. JOHN
Institut für Geologie
Ruhr-Universität Bochum

---

* All figures quoted in the text are at the end of the lecture.

Introduction

The following set of lecture notes is composed of excerpts of previous papers on the given subject, with specific comments added.

The geologic factor of greatest significance in rock mechanics and rock engineering is considered to be the geologic structure represented by joints, faults, and other planes of weakness. This geologic structure differentiates between the **rock element** and the **rock system** composed of these elements.

If rock mechanics is to achieve real significance in engineering practice, it has to provide methods of analysis which are realistic compromises between the best representation of the actual conditions and pragmatic engineering. The rock mechanics practitioner has to face the fact that many geological engineering and rock mechanics problems may be too complex to allow rigorous analysis but at the same time are deemed satisfactory for construction of major structures. Ultimately, the geotechnical engineer or the engineering geologist has to express the effect of geologic factors in quantitative, even if approximate, terms and provide numerical data on strength and deformability of rock systems to be utilized in construction. Qualitative evaluations, quantitative descriptions of geologic features as such, and comparisons of specific test results are always of interest but may not suffice as basis for engineering decisions and designs.

In the following, concepts are given to quantitatively evaluate the effect of simplified geologic structures. In the present approach, it is proposed to express the properties of a rock system in terms of the parameters of both rock elements and the geologic structure.

At this point, reference is made to a most crucial point of any engineering analysis covering geologic materials, the realistic but still manageable idealization of natural conditions, which is to be the input for any analytical attempts. This problem is practically the same for simple and sophisticated methods of analysis of model approaches. Idealization and results based on it always need to be viewed simultaneously in deriving conclusions and recommendations. It is believed, therefore, at the very basic approach given here is justified for engineering purposes in spite of the availability of sophisticated, computer-based methods of analysis.

Engineering problem

The following specific engineering aspects are to be covered in this approach, with the respective parameters to be assessed quantitatively

1.   Strength of regularly jointed rock which is subject to sliding apart upon compression

1.1 Two-dimensional problem as generally considered in engineering studies

1.2 Three-dimensional problems as generally posed due to spatial aspects of the geologic structure

2.   Deformability of regularly jointed rock due to sliding apart

The analytical approach presented here was derived from the results of biaxial compression tests on regular systems of model rock blocks, the resulting concepts are to be applied to the practical problems as sketched below.

In the present approach the properties of the **rock system** (for example a rock abutment supporting an arch dam) is to be expressed in terms of the parameters of the **rock subsystems** and the **major discontinuities** (geologic structure consisting of joints, seams and faults). It is assumed that the properties of the subsystems and of geologic planes of weakness can reliably be assessed by means of in-situ rock tests.

Terms and Graphs

The following terms which are also given in Fig. 5 were deemed convenient in digesting the data obtained. This figure also gives the designation of the applied stresses as used throughout this paper.

Stress ratio is the ratio of compressive vs. confining stresses, with the latter being constant for any test. A test commenced at a ratio of unity for the hydrostatic condition is then increased to the "limiting stress ratio" at the point of failure. Based on these ratios and the confining pressures, intermediate compressive stresses and the final compressive strength can be determined.

System/element coefficients are the ratios correlating the parameters of the system to those of the elements, expressed in percent. These coefficients represent the ultimate results of the analytical process discussed. For the strength, the system/element coefficients are defined as the ratios of either the limiting stress ratios or the compressive strength values, always for the same confining pressures. The system/element coefficients for the deformability are given by ratios of the deformation moduli, again at equal confining pressures and for the identical compressive stress ranges.

The plotting convention used is illustrated in Fig. 6. It shows how the parameters discussed before are graphically related to the orientation of the first set of joints of the model systems by means of polar cooordinates.

Strength of Rock Systems in Plane Conditions

The combination of the Mohr's circle diagram with the Coulomb criterion for the shear failure (in the following Mohr-Coulomb concept) is used to determine the limiting stress ratios for the plane conditions of a section. For a shear failure of an element or of assemblies of elements, without participation of the joints, the conventional soil-mechanics formulation appears to be applicable. For a failure of the system by sliding along one joint or one set of joints, the limiting stress ratios can be determined based on parameters of the joints and the confining pressure. Fig. 7 presents the respective equations.

The limiting stress ratios for a rock system with one family of joints (or one joint only) determined by means of the foregoing relationships are presented in Fig. 8. Three different modes of failure can be differentiated analytically and were verified experimentally.

1) **Sliding along Joints** — For the special case of strictly frictional shear resistance along the joints, the limiting stress ratio is solely dependent upon orientation and friction of the joint. This results in one configuration for one angle of friction when using the plotting convention of Fig. 6. The minima for the limiting stress ratios at the most critical joint orientations can be determined by a term identical to that used in soil mechanics.

2) **Shearing of Elements, Without Participation of the Joints** — The limiting stress ratios depend on the shear parameters of the elements and the confining pressure resulting in a set of concentrical circles for different confining stresses.

3) **Shearing of Elements Affected by Joints** — This mode of failure takes place in the transition zones between the two principal failure modes. The respective limiting stress ratios depend on the parameters of elements, joints, and the confinement.

The data of Fig. 8 can easily be converted to system/element coefficients for strength, resulting in Fig. 9. The reduction in strength of the system, as compared to that of the elements, due to the single set of joints represented, varies with different confining pressures.

For rock systems with several sets of joints, or several individual joints, the graph of Fig. 8 (or of Fig. 9) can be combined by superposition in accordance with the orientation of these joints.

To illustrate the application of above formulations the reduction in strength of a plane system with two joints is determined using the numerical data of

Fig. 10. Fig. 11 represents the results in a polar diagram, with the **directions** representing the orientation of the first joint (or set of joints), which is assumed to be continuous, with an angle of friction of 25 degrees. The second joint of the system, assumed to be discontinuous, with a two-dimensional extent of 50 per cent, and an angle of friction of 40 degrees, intersects the first joint at an angle of 60 degrees. The **amounts** presented in the graph represent the remaining strength of the system with respect to the strength of the element, expressed in terms of the system/elements coefficient f. The effects of different confining conditions are expressed in terms of the coefficient u.

Strength of Rock Systems in Spatial Conditions

A combination of the previous concept with plane projections of the reference hemisphere as used in the geological sciences permits consideration of three-dimensional problems which are of real importance in rock engineering.

The reference hemisphere is a graphic tool to determine the angular relations of planes and directions in space, problems which are somewhat difficult to solve by conventional analytical methods. In the present concept, this tool is used to determine the angle α between the direction of principal compressive stressing of a rock system and the direction of potential sliding along any given geologic plane. This angle enters the equations of Fig. 7. The lower reference hemisphere at the left of Fig. 12 illustrates the three steps necessary to solve this geometric problem, considering again a system with one plane only. From the resulting plot of the angle α vs. the sliding direction, projected into a horizontal plane which is oriented towards north, and the shear parameters of the joint, the limiting stress ratios can be computed and plotted against the direction of sliding as shown on the right side of Fig. 12. By superposition, the effect of several sets of joints can also be evaluated. Relating the final graphic configuration of the limiting stress ratios thus produced with the actual confining conditions, both in respect to confining stress and geometrical freedom of movement, the bearing capacity against "sliding apart" of a discontinuous system can be evaluated quantitatively.

Shear Test of Joints

Fig. 14 presents the hypothetical results of conventional shear tests of continuous flat joints. Two different potential shear behaviors are possible, A and B, which may represent the limits of the actual behavior of either natural or model joints. Fig. 14 gives both the shear stress vs. displacement diagrams for different normal stresses and the equivalent data in Mohr diagrams.

The most significant differences of the two shear behaviors with respect to this study are the configurations of the lines of equal displacement in the Mohr plot. In as much as the model joints performed very much in accordance with the shear behavior B, this concept is based on such relationships; however, adaption to the shear behavior of actual joints is easily possible.

For irregular joint surfaces, the concepts of Fig. 14 can be combined with those of Patton [1] on the shear behavior of toothed model joints. However, in the present study, rock systems with flat joint planes are considered.

The unit joint stiffness as defined by Goodman, both in normal and tangential directions, as ratio of stressing vs. displacement is believed to ultimately be a key parameter for the deformability of discontinuous rock systems. It can also be adapted to discontinuous joints by basing it on parameters of both joints and elements.

Deformability of Rock Systems

The deformations of a discontinuous rock system can be attributed to the rock elements and the geologic structure. The deformability of rock systems prior to failure by sliding along a governing set of joints (also called design joints) was particularly investigated. This mechanism of sliding apart produces rock systems with the greatest reductions in strength and increases in deformability as compared to the rock elements.

In order to derive parameters describing the deformability due to sliding along one set of joints, the following analytical steps are combined.

1) **Mohr-Coulomb concept** (see Fig. 7) to relate the stressing of the system with the stresses in the joint planes.

2) **Results of shear tests of joints** (see Fig. 14B) to relate shear and normal stresses in the joints with the resulting displacements and dilatancies.

3) **Geometrical analysis** (as outlined on Fig. 15) to convert shear displacements and dilatancies at each joint (combined with the joint spacing) to strain of the system due to sliding along the given set of joints.

This "system" permits the determination of moduli of deformation and Poisson's ratios due to **sliding apart**. Acceptance of the shear behavior B of Fig. 14 results in a shortcut in this process inasmuch as intermediate shear displacements at each joint prior to failure can be expressed in terms of intermediate angles of friction. By means of the Mohr-Coulomb concept, the polar plot of Fig. 16 is possible. It is essentially identical to that of Fig. 8 but with the addition of the displacements at any single joint prior to the sliding failure. This graph combines the first two steps of the previously given sequence of analysis, i.e., the conversion of stress ratios to shear displacements.

Adding the geometrical analysis of Fig. 15 and selecting confining pressure and orientation of the joints, the deformation of the system can be computed. In Fig. 17 the deformation of the elements and the deformation due to sliding along the set of joints are combined for one specific example. In this figure, the system/element coefficients for both strength and deformability are given, together with the moduli of deformation of the elements, due to the joints, and of the system. Poisson's ratios, however, are not shown.

The foregoing process can, in principle, also be combined with the reference sphere geometry to at least assess the deformability in three-dimensional problems.

Conclusions

The present thinking of the lecturer on this subject can be summarized as follows

1) When dealing with discontinuous rock systems, strength and deformability represent equally important design criteria.

2) Shear tests on joints provide not only data to determine the stability or the safety factor of rock systems but also essential input to evaluate the deformability of discontinuous rock systems.

3) The concept presented provides a simple analytical basis for the practitioner's estimate of the engineering parameters of in-situ rock, in terms, expressed in percent, of the respective parameter of the integral rock core (rock element). Geotechnical problems encountered in engineering practice have been assessed following this general approach.

4) The concept illustrates the engineering use of parameters describing geologic features, such as orientation, spacing and also extent of joints, which are increasingly provided by engineering geologists but rarely used in analysis performed in the engineering practice. In the time since proposing the present approach it has been proven that the most sophisticated methods of analysis are useless if realistic input data cannot be provided.

5) Since its conception in 1969 the concept has been utilized in developing finite element simulations of jointed rock systems, which are gradually being accepted

by the engineering practice.

6) The model test, which resulted in Poisson ratio μ in excess of 0.5 due to the "sliding-apart" mechanism, provide a basis for the existence of high lateral in-situ stresses not only due to tectonic stresses but also due to the geologic structure.

REFERENCES

[1]   Adler, L., "Failure in Geologic Material Containing Planes of Weakness", Trans. SME/AIME, Vol. 226, 1963, pp. 88-94

[2]   Müller, L., "Der Felsbau", Vol. 1, Enke Stuttgart, Germany, 1963.

[3]   Donath, F.A., "Strength Variation and Deformation Behavior in Anisotropic Rock," State of Stress in the Earth's Crust, W.R. Judd., ed., American Elsevir Publishing Co., New York, 1964, pp. 281-289.

[4]   Krsmanović, D., and Langof, Z., "Large-Scale Laboratory Tests of the Shear Strength of Rock Material", Rock Mechanics and Engineering Geology, Suppl. I, 1964, pp. 20-30.

[5]   Patton, F., "Multiple Modes of Shear Failure in Rock and Related Materials", Ph.D. Thesis, University of Illinois, Urbana, Ill., 1966.

[6]   Sirieys, P.M., "Phénomènes de Rupture Fragile des Roches Isotropes et Anisotropes", Rheology and Soil Mechanics, J. Kravtchenko and P.M. Sirieys, eds., Springer, Berlin, 1966, pp. 396-404.

[7]   Walsh, J.B., "Elasticity of Rock: A Review of Some Recent Theoretical Studies", Rock Mechanics and Engineering Geology, Vol. IV/4, 1966, pp. 283-297.

[8]   Bray, J.W., "A Study of Jointed and Fractured Rock, Parts I and II", Rock Mechanics and Engineering Geology, Vols. V/2-3, V/4, 1967, pp. 117-136, 197-216.

[9]   Goodman, R.E. and Taylor, R.L., "Methods of Analysis for Rock Slopes Abutments: A Review of Recent Developments, Failure and Breakage of Rock", C. Fairhurst, ed. AIME, New York, 1967, Chap. 12, pp. 303-320.

[10] Goodman, R.E., Taylor, R.L. and Brekke, T.L., "A Model for the Mechanics of Jointed Rock", Journal of Soil Mechanics and Foundation Div., Proceedings, American Soc. of Civil Engineers, Vol. 94, SM3, May 1968, pp. 637-659.

[11] John, K.W., "Graphical Stability Analysis of Slopes in Jointed Rock", Journal of Soil Mechanics and Foundation Div., Proceedings, American Soc. of Civil Engineers, Vol. 94, SM2, Mar. 1968, pp. 497-526.

[12] Judd, W.R., "Problems that have arisen during Ten Years of Rock Mechanics", Festschr. 10 Jahre IGB, Akademie-Verlag, Berlin, 1968, pp. 128-151.

[13] Stagg, K.G. and Zinkiewicz, O.C., "Rock Mechanics in Engineering Practice", John Wiley, London, 1968.

[14] Jaeger, J.C. and Cook, N.G.W., "Fundamentals of Rock Mechanics", Methuen, London, 1969.

[15] John, K.W., "Festigkeit und Verformbarkeit von druckfesten, regelmässig gefügten Diskontinuen ", Publication, Institute of Soil Mechanics and Rock Mechanics, Karlsruhe University, Karlsruhe, Germany, No. 37, 1969.

[16] John, K.W., "Civil Engineering Approach to Evaluate Strength and Deformability of Regularly Jointed Rock", Eleventh Symposium on Rock Mechanics, June 1969, AIME, 1970, Chap. 5, Sec. 1.

[17] Kusnezov, G.N., "Graphical Method to Determine the Strength of a Non-Homogeneous Jointed Rock Mass", Rock Mechanics, Vol. 2, No. 2, 1970.

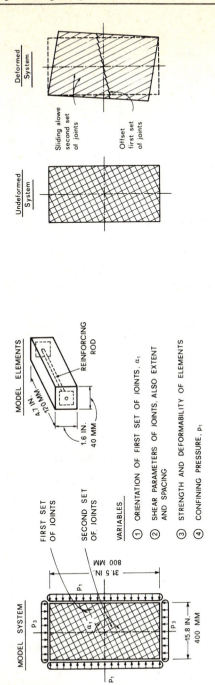

MODEL SYSTEM

FIRST SET OF JOINTS

SECOND SET OF JOINTS

31.5 IN.
800 MM

15.8 IN.
400 MM

P_1 P_3 α_1 P_1 P_3

MODEL ELEMENTS

REINFORCING ROD

4.7 IN.
120 MM

1.6 IN.
40 MM

VARIABLES

① ORIENTATION OF FIRST SET OF JOINTS, α_1

② SHEAR PARAMETERS OF JOINTS, ALSO EXTENT AND SPACING

③ STRENGTH AND DEFORMABILITY OF ELEMENTS

④ CONFINING PRESSURE, p_1

Fig. 1 Model tests of regularly jointed systems.

Deformed System

Undeformed System

Sliding alowe second set of joints

Offset first set of joints

Fig. 2 "Sliding Apart" along one set of joints.

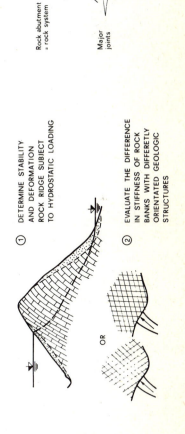

① DETERMINE STABILITY AND DEFORMATION ROCK RIDGE SUBJECT TO HYDROSTATIC LOADING

② EVALUATE THE DIFFERENCE IN STIFFNESS OF ROCK BANKS WITH DIFFERETLY ORIENTATED GEOLOGIC STRUCTURES

OR

Fig. 3 Sample problems.

Arch daw

Rock subsystem

Minor jointing

Rock element

Rock abutment = rock system

Major joints

Fig. 4 Rock element, rock system.

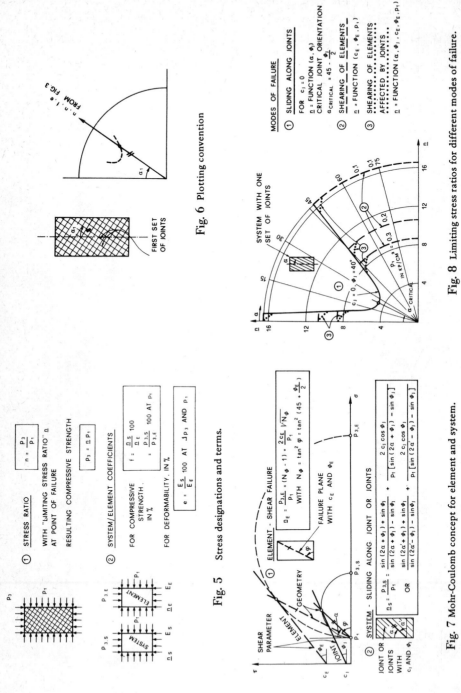

Fig. 6 Plotting convention

Fig. 8 Limiting stress ratios for different modes of failure.

Fig. 5 Stress designations and terms.

Fig. 7 Mohr-Coulomb concept for element and system.

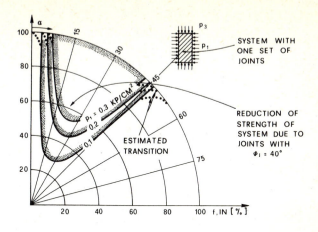

Fig. 9 System/element coefficients for compressive strength

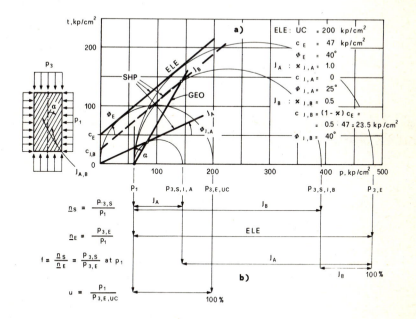

Fig. 10

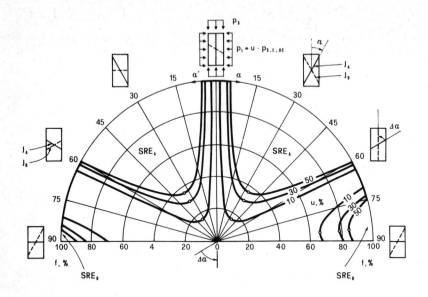

Fig. 11

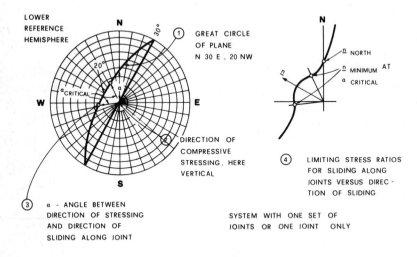

Fig. 12 Spatial evaluation of limiting stress ratios for sliding along one joint.

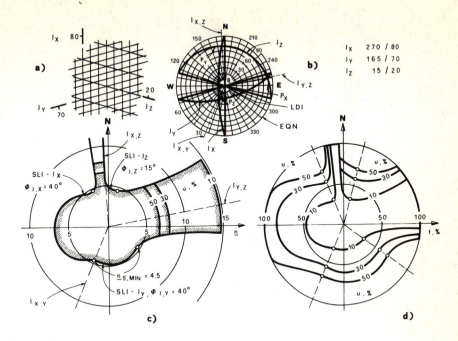

Fig. 13

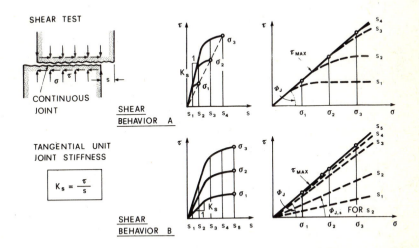

Fig. 14 Hypothetical results of shear tests of continuous joint

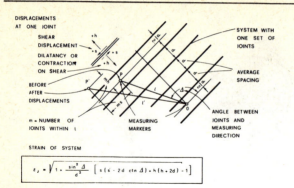

Fig. 15 Strain of system due to shear displacements

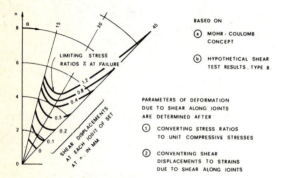

Fig. 16 Stress ratios vs. shear displacements

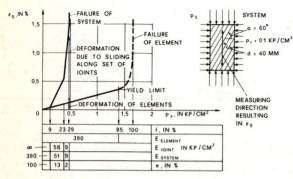

Fig. 17 Computed deformation of system.

FRICTION PROPERTIES AND FRICTIONAL BEHAVIOR
OF ROCK SEPARATION PLANES

N. RENGERS
International Institute for
Aerial Survey and Earth Sciences
(ITC)
Enschede — The Netherlands

---

* All figures quoted in the text are at the end of the lecture.

1. Introduction

Since the birth of rock mechanics as a separate branch of science a few decades ago, it has been clear that the laws of continuum mechanics could not be applied to the mechanical behaviour of rock masses near the earth's surface. The concept of discontinuum mechanics was developed, in which the deformational behaviour of rock masses takes place by relative movements along discontinuities like faults, joints, bedding-planes, etc.. The separation planes may be present before deformation starts or may be caused by fracturing of the rock mass due to local concentration of stresses during the loading process.

The resistance against relative movement along the planes of separation in a rock mass is referred to as friction or shear resistance. Although large amounts of the resistance against relative movement along pre-existing planes can be attributed to shearing resistance of asperities on both sides of the separation planes, we will here use the terminology of **friction** for all processes of relative movement along pre-existing planes and we will use **shear** for relative movement by creation of new discontinuities. The simplest friction criterion is the Coulomb criterion:

$$T \geqslant \mu.N$$
$$\text{or} \quad \tau \geqslant \mu.\sigma$$

in which T is the tangential force necessary for relative movement, N is the normal force and μ is the coefficient of friction which can also be expressed as tan φ (φ being called angle of friction). σ and τ are the stresses in normal and tangential direction, obtained by division of N and T by the area of apparent contact between the two bodies.

2. History of Friction Investigations

LEONARDO DA VINCI (1452–1519) was already puzzled about the nature of frictional resistance and found the proportionality between normal force and friction, as well as the independency of frictional resistance from the area of contact between the sliding bodies. AMONTONS, almost two centuries later, came – ignorant of da Vinci's work on friction – to these same conclusions and explained the nature of friction by assuming a submicroscopic roughness on the surface of both sliding bodies which causes the bodies to slide up and down to interlock again in the next asperities.

EWING (1892) is the first scientist to attribute friction to molecular cohesion across the contact plane.

For metals KRAGELSKI (1965) developed a theory of friction which is based on the deformational work which has to be done in the contact area in the form of plastic deformation or due to hysteresis in elastic bodies. Another theory was developed by BOWDEN and TABOR (1958) who visualize the frictional process as a continuous process of "cold welding" of the isolated point contacts between the bodies and shearing through of these new formed material bridges to enable relative movement. TERZAGHI (1925) used this theory for friction between glass as well as quartz grains. However, investigations by GRIGGS, TURNER and HEARD (1960), BRACE (1963) show that most of the rock-forming minerals behave brittle, even under conditions of extremely high pressure and temperature. Based on this knowledge about the deformational behaviour of rock, BYERLEE (1967) developed a completely new theory for rock friction along flat rock planes which can also be applied to friction along not interlocking rough surfaces. His theory is based on brittle fracture of the microscopic asperities by bedding due to application of tangential stress to the top of the asperity rather than by shearing.

Using the cone model for the asperities he comes to the theoretical value for the factor of frictional resistance μ of approximately 0.1. Friction testing with polished minerals by HORN and DEERE (1962) resulted in friction factors of this order of magnitude.

The above mentioned investigations were based on the frictional behaviour of either flat or non interlocking rough surfaces. In geomechanical practice, however, usually very rough and often interlocking separation planes are met. GOLDSTEIN et al. (1966) and PATTON (1966) show that during relative movement between interlocking rock-blocks a combination of friction and sliding up or shearing through takes place (Fig. 1), with the friction criterion $\tau \geqslant \sigma.tg (\varphi + i)$ in which i is the angle of sliding up. The terminology "of sliding up" is used here instead of "dilatancy" which is often used by other authors. Sliding up being a phenomenon restricted to the process in the separation plane itself, dilatancy being a phenomenon in shear deformation where a **zone** of sheared material increases in volume.

At higher levels of normal stress, the shearing of the interlocked asperities after criterion ③ in Fig. 1 will require less energy then sliding up after criterion ②. This way a bilinear criterion for friction/shear of interlocking roughness results. However, in nature roughness is present in many different forms with different inclination angles i, which will cause a curved criterion. JAEGER (1971) gives an overview of other friction criteria as proposed by different rock friction investigators.

3. Relation Between Surface Geometry and Friction Properties

As shown by GOLDSTEIN et al. (1966) and PATTON (1966) the surface roughness leads to a sliding up in the separation plane which increases the angle of friction with the angle i of sliding-up. Qualitative investigations of these authors using model bodies with simplified surface geometry have shown the validity of this simple relationship between surface geometry and sliding-up behaviour.

The first attempt to compare the results of surface roughness measurements on natural rock separation plot with the results of friction tests of a rather large scale was made at the Rock Mechanics Department of the University of Karlsruhe (RENGERS, 1971). Roughness measurements were taken with stereo-microscope and profilograph and a special way to represent surface geometry was developed for this purpose. (RENGERS (1970), FECKER and RENGERS (1971)).

It could be calculated that the sliding-up-curve during relative movement from the interlocked position has the general form of figure 3a at values of normal stress which favor the sliding up over the shearing (curves②and③in Fig. 1). The extreme steepness of the smallest size asperities gives rise to a high resistance against relative movement due to the great angle i of sliding up. With increasing relative movement the angle of sliding up decreases and with it decreases the effective angle of friction. Thus the typical working diagram tangential force/relative movement for interlocked planes as shown in Fig. 3b is obtained: the initial- or peak value for the frictional resistance is reached at very small amounts of relative movement and decreases to a residual value after a certain amount of relative movement.

4. Methods to Determine Rock Properties

JAEGER (1971) shows the different systems which can be used to determine friction properties in the laboratory and in the field (Fig. 2). Principally the following methods can be distinguished:

a. conventional shear box

b. small gliders on larger surfaces

c. triaxial friction testing on rock cores

d. double shear

e. rotational shear

f. large scale in situ friction testing

All of these testing methods have their special advantages and disadvantages and economy, as well as scales of the samples which can be tested.

Friction testing is either executed with constant normal force or with increasing or decreasing normal force. Tangential force necessary to maintain relative movement as well as amount of sliding-up are plotted against amount of relative movement in the way as shown in Fig. 3.

Results of testing procedures at different levels of normal stress are plotted as single points in a diagram of T against N (or τ against σ) as in Fig. 1.

5. Experimental Work to Determine Influence of Surface Geometry on Frictional Behaviour

At the Rock Mechanics Department of Karlsruhe University a special friction-testing machine was developed with which samples with a testing plane of 40 x 15 cm^2 can be sheared with normal forces and tangential forces up to 50 tons. (RENGERS (1971)).

As the detailed investigation of the sliding-up behaviour during relative movement was one of the most important aspects of the planned experimental work, the constantness of normal load even with rapid and large amounts of sliding up had to be ensured. For this reason air bellows were used for the normal loading, the usually applied hydraulic jacks having the disadvantage of a relatively high stiffness which causes large increases in normal load at small amounts of sliding up.

The material tested in the experimental program was a fresh, fine grained, homogeneous and isotropic granite with high strength from the Black Forest. Separation planes were made by splitting the samples in a direction parallel to the main joint system. With 14 pairs of sample blocks over 100 friction tests were executed for which 4 different arrangements of the blocks were used (see Fig. 4).

a. with and without interlocking of surface geometry.
b. with and without infilling material (which originated during the relative movement by shearing and grinding).

Fig. 5 gives in schematical form the frictional behaviour of four different types of arrangements, which show the characteristics of the behaviour for separation planes in natural conditions.

Arrangement I: Planes with exact interlocking and without infilling are present in nature in the form of tight joints without relative movement. They show a clear peak at small amounts of relative movement which rapidly decreases to a residual value during relative movement.

Arrangement II: Joints without relative movement but with some opening and infilling due to weathering or transported material show a gradual increase of frictional resistance up to the moment of contact between the asperities at both sides of the plane during relative movement.

Arrangement III: This arrangement of non-interlocking surfaces without infilling material is seldom encountered in nature, however it is the usual arrangement in geomechanical model testing where unit rock blocks are assembled in such a way that they resemble a jointed rock mass. Sliding-up and working diagram are completely different from natural joint behaviour, which should be taken into account very well at the interpretation and extrapolation of the result of model testing. The very small angle of initial friction is in agreement with BYERLEE's (1967) calculation based on breaking off of asperities.

Arrangement IV: Non-interlocking planes with fault infilling are comparable with all planes in nature along which relative movement has already taken place. Very remarkable is the fact that, neither in the direction of previous movement (IV A), nor in case of reversal of relative movement (IV B), a peak value for frictional resistance is observed.

During relative movement a combination of sliding up, shearing through and grinding of the material present in the separation plane takes place. RENGERS (1971) shows that an important part of the wearproducts originate during the first relative movement. Grainsize analysis of the infilling material after different amounts of relative movement shows that during relative movement the percentage of fines increases due to fragmentation of the fill material.

In Fig. 6 it is attempted to show the different components of resistance against relative movement from the interlocking position. The angle of residual friction resistance is increased with different angles of sliding-up for different values of normal stress as observed from the sliding-up behaviour during the friction testing. Thus the curved interrupted line gives the total resistance as a combination of friction and sliding up. However, the line $\varphi = 65°$ is the result from the friction testing which leads to the conclusion that the shearing of asperities is responsible for the resistance in the horizontally hatched field. This component clearly increases more than proportionally with increasing normal stress.

The results of the mentioned investigations are applicable only to the investigated rock type, however the author expresses as his opinion that the general conclusions will be applicable to most rock friction processes.

REFERENCES

[1]   Amontons, M. De la résistance causée dans les machines. Mémoires de l'Academie Royale des Sciences, Vol. 1699, 529 (1734).

[2]   Bowden, F.P. and D. Tabor, The friction and lubrication of solids. Oxford: Clarendon Press, Part 1 (1958), Part 2 (1964).

[3]   Byerlee, J.D., Theory of friction based on brittle fracture. Journal of Applied Physics, Vol. 38, Nr. 7, 2928 (1967).

[4]   Ewing, A., Not. Proc. Roy. Instn., Vol. 13, 387 (1892).

[5]   Fecker, E. and N. Rengers, Measurement of large scale roughness of rock planes by means of profilograph and geological compass. Symposium Soc. Internat. Méchanique des Roches, Nancy 1971.

[6]   Goldstein, M., Goosev, B., Pyrogovsky, N., Tulincv, R., and Turovskaya, A., Investigation of mechanical properties of cracked rock. Proc. of the First Congress of the Intern. Soc. for Rock Mech., Lisbon, Vol. 1, 521 (1966).

[7]   Griggs, D.T., F.J. Turner and H.C. Heard, Deformation of rocks at 500° to 800° C. Geological Society of America Memoir, Vol. 79, 39 (1960).

[8]   Horn, H.M. and D.U. Deere, Frictional characteristics of minerals Géotechnique, Vol. 12, 319 (1962).

[9]   Jaeger, J.C., Friction of rocks and stability of rock slopes. Géotechnique, Vol. 21, 97 (1971).

[10]   Kragelski, I.V., Friction and wear. Washington: Butterworths (1965).

[11]   Leonardo da Vinci, Notebooks (translated into English by Edward MacCurdy) London: Jonathan Cape (1938).

[12]   Patton, P.D., Multiple modes of shear failure in rock. Proc. of the First Congress of the Intern Soc. for Rock Mech., Lisbon, Vol. 1, 509 (1966).

[13]   Rengers, N. Unebenheit und Reibungswiderstand von Gesteinstrennflächen. Dissertation Karlsruhe (1971). Veröffentlichungen des Institutes für Bodenmechanik und Felsmechanik, Karlsruhe.

[14]   Terzaghi, K., Erdbaumechanik, Leipzig und Wien: Franz Deuticke (1925).

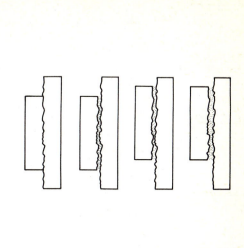

Fig. 2 Different systems used to determine friction properties (after JAEGER).

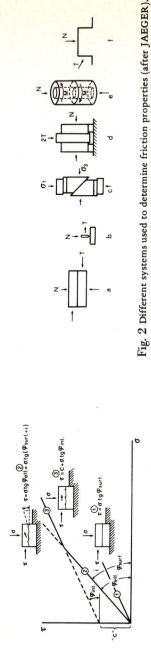

Fig 1   Sliding up and shearing through of interlocked asperities
        (after GOLDSTEIN et al. and PATTON)

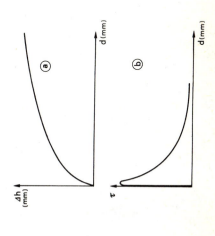

Fig. 3 Sliding-up curve and working diagram for relative movement along interlocked, rough separation planes in rock.

Fig. 4 Different arrangements for friction testing, with and without infilling material   and with and without interlocking of roughness.

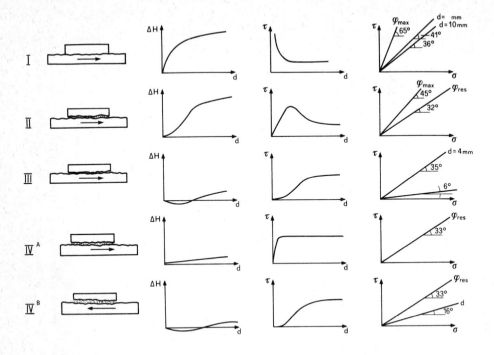

Fig. 5 Frictional behaviour of different types of test arrangements.

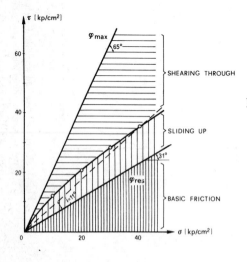

Fig. 6 Relationship between sliding-up shearing through, and basic friction for Malsburg granite at different normal stresses.

ANALYTICAL METHODS FOR ROCK SLOPE ANALYSIS

H.K. KUTTER
Institut für Geologie
Ruhr-Universität Bochum

---

* All figures quoted in the text are at the end of the lecture.

Introduction

Although a rock slope is not normally thought of as an engineering structure in the conventional meaning of the word — probably because it is either a natural slope or, if man-made, a kind of negative structure built by taking material away rather than by assembling it together — its design follows basically the same principles and requirements as that of any other structure. The solution of any structural problem requires first the definition, knowledge or assumption of five essential factors:

1) The geometry of the entire structure i.e. its external boundaries and its internal discontinuities (e.g. geological boundaries).
2) The external loads and body forces (static and dynamic) and force histories to which the structure is exposed.
3) The relevant mechanical properties of the materials which constitute the structure.
4) A definition and mechanism of failure and a criterion for its initiation or progression, which all are highly dependent on the type of structure and the purpose which the structure has to serve.
5) The factor of safety or the acceptable probability of failure based on past design experience and in effect a factor of overdesign necessary because of insufficient knowledge of any of the four previous factors.

Before going into a discussion of the techniques available for the analytical solution of slope stability problems, a few comments and clarifications on the above listed design factors are first needed. Rock is in the rarest of cases a continuous uniform material. It is full of discontinuities and weak structural features which greatly control the strength of the entire rock mass and consequently also the failure surfaces. The geometry of the potential sliding mass of an unstable rock slope is therefore determined by the location and orientation of the discontinuities. Only in the rare case of randomly oriented and relatively narrow-spaced joints will the rock mass behave similarly to a soil where the failure surface is curved and determined by the minimum energy principle. In the more normal case, however, the potential failure surfaces are dictated by the structural weaknesses, such as fault zones, joints, bedding and foliation planes. A thorough site investigation and a careful mapping of the discontinuities are therefore the first and most important step in a slope stability analysis. Dependent on the orientation, spacing and distribution of discontinuities the types of rock slope failures due to sliding can be categorized as illustrated in Figure 1. The complexity of the slide geometry generally increases with the volume of the sliding mass and the initial or triggering failure may be of a very different kind than the final one. In any stability analysis a number of likely failure geometries should therefore be investigated.

The static loads and the body forces acting on the sliding mass are, with the exception of the water pressure, in most cases clearly defined. The weight of the sliding mass stands of course in direct relation to the failure geometry. In a dry slope few uncertainties derive therefore from the load assumptions. In a wet slope, where the water pressure can contribute a substantial percentage of the total forces, the generally insufficiently known water pressure distribution introduces a big factor of uncertainty into the analysis.

The relevant mechanical properties in a slope analysis are predominantly the strength parameters of the discontinuities. These are discussed in a separate lecture and the question as to the proper choice of relevant parameters is best dealt with in the later paragraph on design methods.

In addition to data on shape and properties of a structure information or assumptions on the most likely mode or mechanism of failure are necessary for a structural analysis. Gravitational sliding of a rock mass is the most frequent type of failure of rock slopes. Another mode, the free fall of rock is generally limited to small moving volumes and to extremely steep slopes, and the not too rare modes of buckling and toppling were usually neglected. They are, however, discussed in another lecture. It is not sufficient to establish only a mode of failure, such as sliding, in addition criteria have to be set up which state the conditions at which failure will start or rupture occur. These conditions can be expressed in terms of stresses, strains, deformations or energy. Stress and strain criteria are very common in structural design. But their application to stability calculations of slopes is certainly very difficult because of the non-uniform stress distribution, the structural discontinuities and the lack of geometrical symmetry. Even modern numerical analysis methods have shown that this classical design approach is not suited for slopes. An exact stress-displacement analysis of slopes would require a representative constitutive equation, the ability to deal with non-linear material behaviour, variable body forces and non-homogeneity, and a knowledge of the initial stresses.

This would lead to too complex a design which simply is not justified by the as yet still poor quality of the input data. However, a simple and semi-empirical, but powerful method, the method of limit equilibrium was found to be particularly useful for slope stability analysis.

The Limit Equilibrium Method
The Technique

This method is widely used for the design of foundations and other surface structures in rock and it is based on the assumption that failure occurs by a sliding movement of a rock mass along pre-existing surfaces of weakness. The location and shape of the failure surface is therefore postulated at the outset rather than being the result of the analysis. The failure criterion is satisfied when at the slip surface a state of limiting equilibrium is reached between the driving forces due to gravity, water pressure and external forces, and the shearing resistance. Since kinematics of the failing rock mass is ignored the slip surface can take any shape. A knowledge of the shear strength of the discontinuities is required, and pore or water pressure, anisotropy and non-homogeneity can be considered. The design procedure consists of finding the critical slip surface for which the factor of safety, which is defined as the ratio of shearing resitance available to shearing resitance required for equilibrium, is a minimum. For a stable slope this minimum value must be larger than one; how much larger than one the factor of safety should be depends on the individual and local conditions, the purpose of the structure and on past experience. One of the limitations of the method of limiting equilibrium is the above mentioned disregard of the deformation and strength of the sliding mass itself; other short-comings are the necessary simplification of the normal stress distribution along the failure surface, the neglect of strength variation with displacement and the limitation of failure modes to shearing only.

Types of limit equilibrium design

All methods and techniques which are currently used in practical slope design are variations of the limiting equilibrium method. The differences lie principally in the geometry of the slip surface and the detail to which the balance of forces or moments is set up. The earlier versions are based on the linear Coulomb criterion

$$\tau = c + \sigma \tan \phi$$

and most of them were developed for soil slopes.

The simplest case is that of planes sliding of a rigid wedge. With the geometry as illustrated and following the earlier definition in Figure 2 the factor of safety for dry conditions is given by

$$F = \frac{W \cdot \cos \alpha \tan \phi + c \cdot 1}{W \sin \alpha}$$

It is assumed that both normal and shear stresses are uniformly distributed over the slip plane. The weight W of the wedge of unit width is

$$W = \frac{1}{2} \gamma H^2 \frac{\sin(i - \alpha)}{\sin i \sin \alpha}$$

and therefore

$$F = \frac{\tan \phi}{\tan \alpha} + \frac{2c \cdot \sin i}{\gamma H \sin (i - \alpha) \sin \alpha}$$

For cohesionless material the expression simplifies to

$$F = \frac{\tan \phi}{\tan \alpha}$$

and the factor of safety is then independent of slope height and wedge shape.

With the presence of groundwater and pore pressure in the slope, the concept of effective stress (= normal stress due to weight minus water pressure) has to be used. For the extreme case of a just fully submerged slope the factor of safety is given by

$$F = \tan \phi \left(\frac{1}{\tan \alpha} - \frac{\gamma_w}{\gamma} \frac{\sin i}{\sin \alpha \sin (i - \alpha)} \right) + \frac{2c \sin i}{\gamma H \sin (i - \alpha) \sin \alpha}$$

where γ_w is the specific weight of water.

To demonstrate the enormous influence of water pressure in a slope a simple example is given. With $\phi = 35°$, $\alpha = 19^0$ and $i = 70°$ the factor of safety for the dry slope is 2.0; however in the case of the just submerged slope it is reduced to 1.0.

In soils the most frequently found shape of slip surface is the circular arc. Here again no distortion of the sliding mass occurs and consequently the analysis should provide an exact solution. In this case the equilibrium equation is written for the moments around the center of the arc [4, 13]. The normal stress at each point of the slip surface is estimated to be equal to the weight of the column of rock or soil which has a unit area cross section and lies vertically above the point. The subsequent graphical analysis, which consists basically of drawing force diagrams is rather lengthy. Simplified procedures were developed which assume simpler stress distributions and use the friction circle method [13, 5]. Their use is however limited to cases where ϕ is the same for the entire slip surface.

In order to be able to analyze slopes with failure surfaces of any shape, with variable ϕ and with non-linear laws of friction the method of slices was developed. The potential sliding mass is divided into vertical slices and the equilibrium of each slice is considered [1]. Vertical and horizontal forces are assumed to be transmitted between the slices. The latest refinements of the method, which only have become practical with the use of modern computers, center on the details which define the position and line of action of these forces [11]. Iterative techniques are applied for the solution of the set of equations which are derived from the equilibrium conditions for each slice. The program developed by Morgenstern and Price can successively deal with complex slope geometries, almost arbitrary distributions of shear strength, density and water pressure [10]. Although these methods have been primarily developed for soil slopes they can be equally well used for the analysis of complex rock slopes.

Application of the Technique to Rock Slopes

Various methods of analysis are available and need practically no modification when applied to rock slopes instead of to soil slopes for which they were originally derived. But before using the most sophisticated analysis one first has to consider the accuracy which is required and the time which is available for the design, and most of all the detail to which the joint patterns and discontinuities are known. Insufficient knowledge of the failure surface hardly justifies a very detailed computation. A decision has to be made therefore whether the use of a simplified and approximate method would be more appropriate, (e.g. for a preliminary design), or whether a detailed analysis (e.g. for the final design) would be required. Accordingly, the approach to a solution should be either via design charts or graphical solutions or hand calculations or computer programs.

A number of rock slope problems can be reduced to the two--dimensional case. Since discontinuities determine the slope surface the majority of failure surfaces in rock are planar or multiplanar. The typical variations of slide geometry are illustrated in Figure 1. The simple case of a single plane of sliding has already been discussed. What remains to be noted is that for the friction-only case the development of a tension crack has no effect on the stability of the slope unless water is present in the rock mass. In the more general case the tension crack must be considered in the analysis since particularly a water-filled tension gap can drastically reduce the stability of a slope.

A single continuous plane of sliding is relatively rare; more frequently the slip surface is formed by two or possibly more sets of conjugate joints, or by one intermittent joint set where the continuous failure surface is formed either by shearing through the so-called rock bridges in a direction parallel to the joints or by tensile failure normal to them. The situation of two conjugate sets forming the sliding surface (Fig. 1, case IIb) has been discussed in detail by Jaeger in the Eleventh Rankine Lecture [7]. With two joint sets, one inclined at the angle α to the horizontal and the other at the steeper angle α , forming a slip surface inclined at the angle β to the horizontal a slope of height H and angle i has a factor of safety

$$F = \tan \phi \, \cot \alpha + \frac{2c \, \sin i \, \sin(\alpha_1 - \beta)}{\gamma \, H \sin(i - \beta) \sin(\alpha_1 - \alpha) \sin \alpha}$$

if the Coulomb law is assumed.

Although it is very difficult to measure or even to observe a regular extent of jointing within an intermittent joint set in the field, a theoretical slope stability analysis for the situation IIc in Figure 1 is possible. A range of most probable two-dimensional extents of jointing should be considered in the analysis for proper assessment of the sensitivity of the design to this structural feature. The extent of jointing k gives that percentage of the slip surface which comprises pre-existing discontinuities and whose shear resistance is therefore that of a joint. The remaining percentage has a shear resistance equal to that of the solid rock. i.e. generally the same friction angle ϕ as the joint, but with a considerably higher cohesion intercept. This case has been discussed by Jennings [8] whose analysis gives for the Coulomb law and dry slope a factor of safety of

$$F = k \, \frac{\tan \phi_j}{\tan \alpha} + \frac{[\, 2(1 - k) S_m + 2kc_j \,] \sin i}{\gamma \, H \sin(i - \alpha) \sin \alpha}$$

where S_m is the shear strength of the intact rock and c_j the cohesion joint.

Instead of shear through solid rock bridges the fractures connecting discontinuous parallel joints can be of the tensile type and normal to the joint system (type IId in Figure 1) resulting in a stepped slip surface. With the intermittent joints inclined at α degrees to the horizontal, the tensile cracks normal to them and an average slip surface inclined at β degrees, the factor of safety becomes [7]

$$F = \tan \phi \cot \alpha + \frac{2\left[c\ \cos(\beta - \alpha) + T_o \sin(\beta - \alpha)\right] \sin i}{\gamma\ H \sin(i - \beta) \sin \alpha}$$

where T_o is the tensile strength of the solid rock.

Whereas the previously discussed two-dimensional rock slope geometries require no distortion of the sliding rock mass, the breaking-up of the block system at the sharp bends of multiplanar slide surfaces (Figure 1, III) together with block rotation, opening of fractures and mass dilatation greatly complicate an accurate analysis. A relatively conservative assessment of stability is however obtained by disregarding the distortion at the bends: divide the sliding mass into slices, with the "cuts" at the intersections of the slip planes, and perform a regular limiting equilibrium analysis allowing forces to be transmitted between slices. Jennings [9] goes through the steps of the analysis of this type of slide geometry.

Finally the point should be stressed that hardly any slope failure geometry is truely two-dimensional. The simplicity of the two-dimensional case, however, frequently justifies the inaccuracies one introduces by analyzing three--dimensional slip surfaces as quasi-two-dimensional ones. The only solutions available for a three-dimensional slide are those for the wedge sliding on a pair of intersecting planes. The methods of solution are either graphical using stereographic projection or analytical in terms of vectors and lately in terms of scalars. Since these methods will be discussed in separate lectures there is no need to describe them here. Only a general comment regarding the third dimension may be added: It is common experience that slopes which are convex in plan tend to be less stable than straight ones, whereas concave slopes are more favourable to stability. The main reason may be a lack of side confinement in the first case and increased confinement and arch action in the latter. Hoeck [6] suggests that "curvature of the slope, in plan, can result in differences of approximately 5 degrees in the critical slope angle".

Rapid Design Methods

In many instants the engineer is faced with the necessity to give an immediate estimate of the stability of a rock slope without having access to complicated computer programs and without sufficient time for a detailed analysis.

At the preliminary stage in the design the various parameters are not or only very poorly known and a type of sensitivity analysis is called for in order to determine the most critical parameters. Refined stability calculations are not very suited for this purpose, but simple and approximate methods should be applied. Two examples for this very practical approach shall be discussed here.

a) Circular failure analysis by Bray [3]

Bray proposed a graphical method for finding the upper and lower bounds to friction angles which are required for limiting equilibrium of a circular arc failure. The determination of upper and lower bound is based on the best and worst normal stress distribution along the slip surface. A stress distribution represented by a single point reaction brings the smallest resisting moment due to friction, whereas a two point stress distribution at the two end points of the slip circle yields the largest resisting moment (Figure 3). The true stress distribution causes a resisting moment which lies between these two limits. The upper bound solution to the limiting equilibrium value of ϕ is therefore given by the latter stress distribution and the lower bound by the first one.

The construction of Bray's upper and lower bound is illustrated in Figure 4. O is the centre of the circular slip arc AB. The resultant force R, composed of the weight of the sliding mass, the total cohesive force, the normal water pressure (uplift) and the horizontal water pressure, acts at G, the centroid of the sliding mass, and intersects the upper part of the circle OAB at T and the slip circle at S. The lower bound solution to the required friction angle is then given by the angle OST. The lines TA and TB are the lines of reaction at point A, respectively B. The upper bound solution is therefore given by the two equal angles OAT and OBT. Bray found that the so obtained bounds are close enough together to be meaningful and useful for practical design.

b) Hoek's design charts for plane and circular failure [6].

The working conditions on site and the speed with which preliminary answers often must be obtained do not favour lengthy calculations. For this reason design charts are developed in many branches of engineering. They present the analytically derived relationships between the relevant variables or dimensionless products thereof and the required solution. In the case of slope stability charts the solution is expressed in terms of the factor of safety and one of the parameters is generally the dimensionaless term $\gamma H/c$, with H the slope height,

γ the rock density and c the cohesion.

Hoek gives charts for the design of rock slopes with circular and planar failure surfaces for a variety of typical cases with and without a vertical tension crack and with different water tables. He found that the results can be adequately obtained by using the two dimensionless functions X and Y, where X was chosen to provide a close approximation to the theoretically derived ideal function. Figures 5 a) and 6 a) illustrate the typical geometries considered and give the respective slope angle functions X and slope height functions Y. The actual design charts are shown in Figures 5 b) and 6 b).

One must be aware that these charts deal only with a few simple slope geometries and do not provide absolute values. But they cover the most important situations which can arise in the cases of circular or planar failure in a rock slope. As long as one keeps in mind the approximations and assumptions which were made in the derivation of these simple charts they can serve a very useful purpose and represent a very valuable engineering tool.

Conclusions

The available literature on rock slope stability analysis is enormous in numbers (12). A two hour lecture cannot deal with the complete field of analytical solutions to this problem. An attempt was made therefore to present only the main ideas and few specific examples. For detailed understanding of the subject a thorough study of the relevant publications is recommended.

REFERENCES

[1]   Bishop, A.W., "The use of the stability circle in the stability analysis of earth slopes", Géotechnique, Vol. 5, No. 1, 1955, pp. 7-17.

[2]   Bishop, A.W. and Morgenstern, N.R., "Stability coefficients for earth slopes", Géotechnique, Vol. 10, No. 4, 1960, pp. 129-150.

[3]   Bray, J.W., Personal communication, 1972.

[4]   Fellenius, W., "Erdstatische Berechnungen mit Reibung und Kohäsion", Ernst Verlag, Berlin, 1927.

[5]   Fröhlich, O.K., "On the danger of sliding of the upstream embankment of an earth dam", Trans. 4th Congress on Large Dams, Vol. 1, 1951, p. 329.

[6]   Hoek, E., "Estimating the stability of excavated slopes in Open Cast Mines", Trans. Inst. Min Metal. London, Vol. 79, 1970, pp. A109-A120.

[7]   Jaeger, J.C., "Friction of rocks and stability of rock slopes", Eleventh Rankine Lecture, Géotechnique, Vol. 21, No. 2, 1971, pp. 97-134.

[8]   Jennings, J.E. and Robertson, A.M., "The stability of slopes cut into natural rock", Proc. 7th Intl. Conf. Soil Mech. Found. Eng., Mexico 1969, Vol. 2, pp. 585-590.

[9]   Jennings, J.E., "A mathematical theory for the calculation of the stability of slopes in open cast mines", Proc. Symp. on Planning Open Pit Mines, Johannesburg 1970, pp. 87-102.

[10]   Morgenstern, N.R., "Ultimate behaviour of rock structures", Chapter 10 in 'Rock Mechanics in Engineering Practice', ed. Stagg and Zinkiewicz, Wiley, London, 1968.

[11]   Morgenstern, N.R. and Price, V.E., "The analysis of the stability of general slip surfaces", Géotechnique, Vol. 15, 1965, pp. 79-93.

[12]   Rock Mechanics Information Service, "Bibliography on slope stability", Rock Mechanics Abstracts, Vol. 1, No. 3, 1970, pp. 269-296.

[13]   Taylor, D.W., "Stability of earth slopes", J. Boston Soc. Civ. Engrs., Vol. 24, 1937, pp. 337-386.

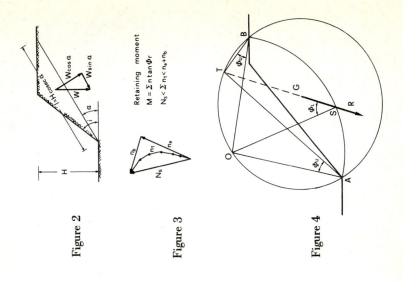

Figure 2

Figure 3

Figure 4

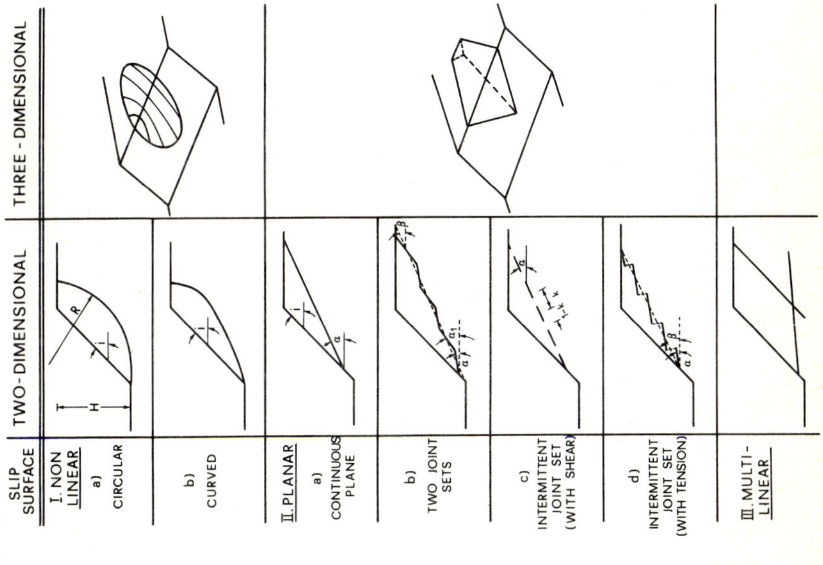

Figure 1

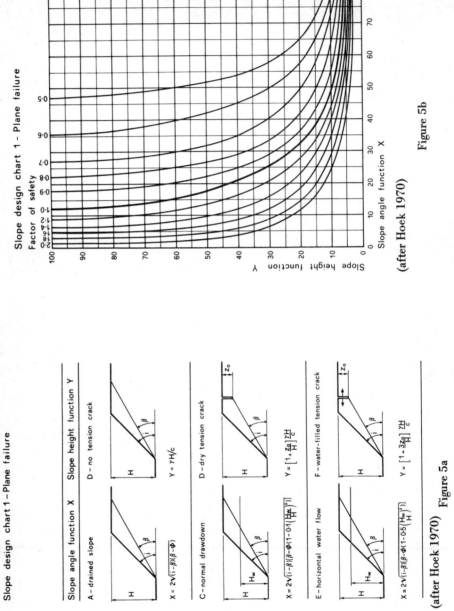

(after Hoek 1970)

Figure 5b

Figure 5a

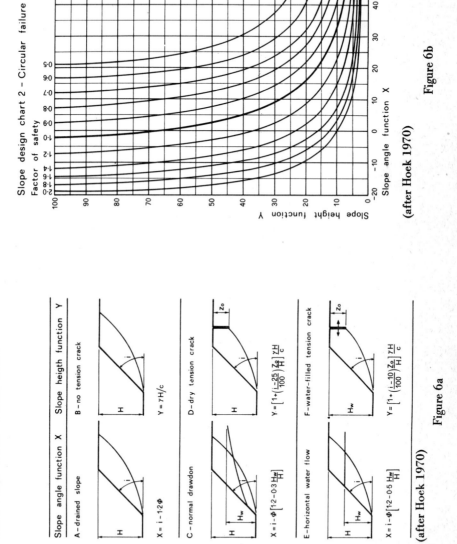

Slope design chart 2 – Circular failure

Slope angle function X   **Slope heigth function Y**

A – drained slope   B – no tension crack

$$X = i - 1.2\,\phi$$ $$Y = \gamma H/c$$

C – normal drawdon   D – dry tension crack

$$X = i - \phi\left[1.2 - 0.3\dfrac{H_w}{H}\right]$$ $$Y = \left[1 + \left(\dfrac{i-25}{100}\right)\dfrac{Z_o}{H}\right]\dfrac{\gamma H}{c}$$

E – horizontal water flow F – water-filled tension crack

$$X = i - \phi\left[1.2 - 0.5\dfrac{H_w}{H}\right]$$ $$Y = \left[1 + \left(\dfrac{i-10}{100}\right)\dfrac{Z_o}{H}\right]\dfrac{\gamma H}{c}$$

(after Hoek 1970)

Figure 6a

Slope design chart 2 – Circular failure

Factor of safety

(after Hoek 1970)

Figure 6b

MECHANISMS OF SLOPE FAILURE
OTHER THAN PURE SLIDING

H.K. KUTTER
Institut für Geologie
Ruhr-Universität Bochum

---

* All figures quoted in the text are at the end of the lecture.

Introduction

Surface rock engineering relies greatly on the theories and techniques developed for soils. Frequently the principles of soil mechanics and the experience gained with soil structures are directly applied to rock slopes and to foundations on rock. This approach is sensible since many aspects and phenomena are the same for soils as for rocks, only that the various parameters are different in magnitude and importance. It is particularly valid when the rock mass is heavily fractured and the joints randomly oriented. However, as soon as directional structural features such as fault and bedding planes or columnar jointing are present in the rock mass the mechanism of failure will be very different from that of a granular medium. A good example is the failure surface of an unstable slope: in soils it is generally a circular arc determined by the minimum energy principle, whereas in rocks it is a preexisting plane or surface of weakness.

Due to the initial student-teacher relationship between rock and soil mechanics practically all stability analyses of rock slopes were run according to the same concept as those for soil slopes, namely that of translational sliding of a rigid mass along an inclined surface. This may still be the predominant mode of failure even for rock slopes, but the presence of regular sets of discontinuities in the rock mass makes other modes of failure possible which have so far been ignored in most slope designs. Rotation of individual rock blocks, subsequent dilation and compaction combined with intermittent slip movements lead to a very different deformational behaviour and consequently to failure modes other than pure sliding. Spalling or buckling of slope faces can occur if the rock mass is higly anisotropic and if for instance zero tensile strength exists in one direction due to poor cementation or high schistosity.

Two types of failure which can only occur in rock slopes with particular joint orientation and distribution shall be discussed here: The phenomena of toppling (= rotation about pivot points) and buckling (= instability of columnar and sheeted rock structures).

Toppling in Rock Slopes

Although the phenomenon of toppling has been previously observed at steep rock slopes, it was generally looked at as the cause for the free fall of a relatively small volume of loose rock blocks rather than a mechanism which can lead to the failure of an entire slope. It was Bray [2] who directed attention to the latter and who initiated the work of Ashby [1] which is the first attempt to analyze this type of failure. Ashby used two-dimensional block models to investigate the relationship between angle of dip, slope angle and column height. The results were compared with those obtained from a computer model [3]. The following paragraphs are based on this study.

Toppling is the rotation of a system of parallel steeply dipping and overhanging slim columns or thin sheets about pivot points. Rotation of the blocks leads to initial slip and the subsequent generation of a continuous slip surface (see Fig. 1).

Toppling is therefore the initial mechanism which has to take place before a slope with discontinuities, which dip steeply into the slope, can become unstable.

The basic concept of toppling is best demonstrated on the case of a single block resting on an inclined base. From the geometry, as illustrated in Figure 2 it becomes clear that the block will topple as soon as the weight vector does not pass through the base of the block, i.e. the breadth-height ratio, b/h, of dip of the base, α . Once α exceeds ϕ, the angle of friction, pure toppling will stop and sliding or a combination of sliding and toppling will occur.

However, an overhanging slope is not necessarily the condition for a toppling failure. Figure 3 shows a situation where the actual slope is not extremely steep but where nevertheless failure is induced by the mechanism of toppling. The development of failure by toppling succeeded by sliding can easily be demonstrated with a "base friction model". This analogue to gravitational loading of a two-dimensional model consists of a horizontally arranged block model with bottom frame which move relative to its base. The frictional forces between base and blocks are analogous to the gravitational body forces. In relation to the model the base moves in the direction of gravity. Figure 3 shows various failure stages of a base friction model of a toppling slope. Three distinct zones of behaviour can be differentiated [1]:

"a) A region of sliding generally restricted to the toe block
 b) A region of toppling columns and blocks with step failure by sliding when the dip of the low angle discontinuities exceed the friction angle of the blocks...
 c) An approximately triangular stable region in which no movement occurs..."

The most significant result of these model tests is the fact that failure of a highly discontinuous rock slope with an inclined through-going discontinuity occurs at a dip angle of this discontinuity which is considerably smaller than the angle of friction (in Ashby's experiments down to 50%). This means that the effective frictional resistance along an inclined plane decreases with an increasing degree of jointing of the potential sliding mass. For practical slope design this is of great significance.

Factors other than those determining the geometry and orientation of the columns affecting the mechanism of toppling are:

a) The degree of jointing or the rock columns.
   Continuous columns fail at angles approaching ϕ, whereas highly jointed columns fail at considerably lower angles.
b) The stiffness of the column base.
   Columns based on a soft layer are less stable than those on hard ones.
c) The frictional characteristics of the through-going discontinuity at the toe.
   Since in most cases toppling only becomes possible when the toe blocks are free to slide, their sliding resistance becomes a critical parameter.
d) Interlocking of the blocks.
   Intimate interlocking of the rock blocks partially prevents their rotation and therefore reduces the danger of toppling.
e) Dynamic effects.
   Dynamic disturbances like blasting vibrations and earthquake tremors accelerate failure due to temporary reduction of the effective frictional forces and addition of unfavourable dynamic forces. A loosening of the block structure favours block rotation.

The formation of a tension crack at the crest of a failing slope can serve as a useful warning device, more so for toppling than for translational slides. In toppling the tension crack is clearly generated before sliding starts at the toe and considerably before rotation of the columns proceeds.

Joints and joint sets which are steeply dipping into the slope were previously considered to play no role in the stability of a rock slope. However, when toppling is taken into account they can lead to failure and are therefore critical to the safety of a slope.

Buckling of Rock Slopes

The failure mode of buckling can only occur in slender structures, i.e. when one or two dimensions of the structural member are considerably smaller than the remaining one(s). Furthermore, the critical force must act in a direction normal to that of the small dimension of the structure. These conditions are fullfilled in columnar or sheeted rock masses with their major joint set parallel to the slope face and their smallest dimension in a direction normal to it. If the sheet or column at the slope face are not free of confinement, i.e. if the discontinuities dip at a steeper angle than the slope face, buckling is not likely to occur.

Little attention was so far paid to the mechanism of buckling in connection with rock slopes. Only recently Walton [4] observed on one of the National Coal Board's opencast mines in Great Britain, how a massive slab slide was actually triggered and preceeded by a buckling failure of the top slab. Starting with a slowly increasing bulging movement in the lower part of the slope face, the instability developed gradually and led to the generation of tension cracks and uplift of the top layer in the critical region. Failure finally occurred after a period of about two weeks in the form of an extensive slab slide.

There is to my knowledge no record of any attempt to analyze this type of slope failure. With a gross simplification of the load distribution and the boundary conditions, however, a rough estimate of the critical length of the top layer can be obtained. Taking friction (ϕ) between the top layer and its base layer into consideration the buckling load may be approximated by

$$P = L \cdot t \cdot \gamma \cos \alpha \, (tg\,\alpha - tg\,\phi)$$

for a slope geometry as illustrated in Figure 4. For a sturt with end conditions as shown in Figure 5 and similar to those observed in the field, Euler's theory gives the buckling load as

$$P_{cr} = \frac{\pi^2 \, E \, I}{(0.7 \, L)^2}$$

where E is Young's modulus and I the moment of inertia. Substituting the above load into Euler's formula the critical lenght for buckling follows to be

$$L = \left[\frac{\pi^2 \, E \cdot t^2}{6 \, \gamma \, \cos \alpha \, (tg\,\alpha - tg\,\phi)} \right]^{1/3}$$

The confining action of the weight component normal to the discontinuity has been completely neglected here. A more complicated analysis taking the true linearly increasing normal load and the lateral load components into account, but based on the geometry of a pin-ended strut gives surprisingly exactly the same answer for the critical length.

A numerical example shall demonstrate the order of magnitude of length required for buckling failure:

$$With \quad E = 35 \cdot 10^3 \ MN/m^2$$
$$\gamma = 25.3 \ kN/m^3$$
$$t = 0.60 \ m$$
$$\alpha = 35°$$
$$and \quad \phi = 30°$$

the critical length calculated is L = 39 m. If friction between the layers is neglected the critical length decreases to only 22 m. From this example it becomes obvious that even in relatively small slopes buckling can be a real danger to the stability.

Finally, a number of factors other than those determining the geometry of the buckling slab or column are discussed:

a) Similarly to toppling, an increase in flexibility of the slab in the form of cross-joints greatly decreases its stability.

b) Ripples and pronounced waveness of the discontinuities lead to deflections of the top layer after a small translational sliding movement and consequently to favourable conditions for buckling.

c) Poor cementation between layers and preexisting shear movements make bukling more likely.

d) Joint water between relatively impervious layers aggravates the safety of a slope prone to buckling.

e) An accelerating failure situation may occur when the top layer of a slope buckles, where the layers dip at a steeper angle than the slope. Each following layer is longer and therefore less stable than the preceeding one as soon as the confining action of the top layer is removed.

Conclusions

It has been described how in jointed rock slopes other failure modes can trigger or preceed that of sliding. The predominant and essential factor which determines the possibility for these types of failure is the geological structure and its orientation with respect to the slope face. In order to assess the stability of a rock slope against buckling and toppling failure a thorough site investigation and mapping of the discontinuities would be the first and most important requirement. Rock slope design without detailed knowledge of the local structural geology should therefore never take place.

REFERENCES

[1]   Ashby, J.P., "Sliding and toppling modes of failure in models and jointed rock slopes", M. Sc. Thesis, University of London, Imperial College Research Report No. T. 3., November 1971.

[2]   Bray, J.W., "A study of jointed and fractured rock", Rock Mechanics and Engineering Geology, Vol. 5, No. 2-3, 1967, pp. 117-136 and 197-216.

[3]   Cundall, P.A., "A computer model for simulating progressive, large--scale movements in blocky rock systems", Proceedings of ISRM Symposium on 'Rock Fracture', Nancy 1971, Paper II-8.

[4]   Walton, G., Geologist with National Coal Board Opencast Executive and member of Rock Mechanics Group, Imperial College London, Personal communication, 1972.

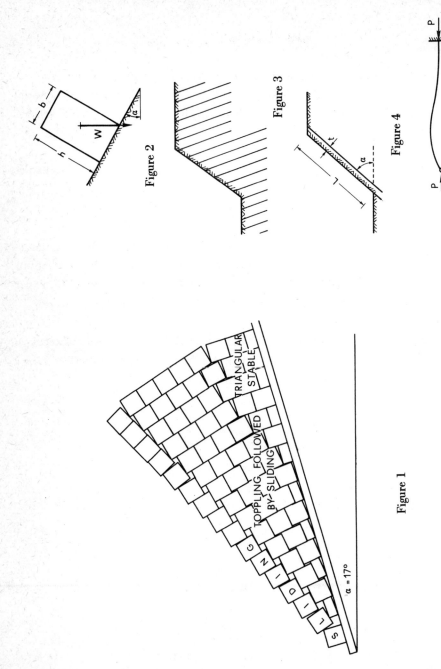

Figure 2

Figure 3

Figure 4

Figure 1

GRAPHICAL METHODS FOR
SLOPE STABILITY ANALYSIS

K.W. JOHN
Institut für Geologie
Ruhr-Universität Bochum

All figures quoted in the text are at the end of the lecture.

Introduction

Nature provided most rock masses with a three-dimensional structure of more or less planar character. This geologic structure differentiates between the **rock element** and the **rock system** composed of these elements. During the last decade it has been firmly established that the geologic structure is, in many cases, of decisive importance for the stability of man-made structures supported on or placed in rock. Conventional geotechnical stability analysis generally consider the sliding of rigid bodies along existing or newly formed planes of weakness in two dimensions only. Three-dimensional problems can be solved by means of analytical geometry, however, these approaches are often too sophisticated for the stability problems encountered in engineering practice.

This lecture covers the use of the hemispheric projection, generally considered to be a tool in mineralogy and structural geology, in combination with conventional engineering methods for the analysis of three-dimensional slope stability problems in jointed rock.

This set of lecture notes mainly consists of excerpts of three previous papers by this lecturer supplemented by an up-to-date listing of references on this specific topic. It should be noted that the material presented in this lecture is intended to serve as first introduction to the proposed approach. For further study specific reference is made to the recent paper by Hoek, Bray and Boyd.

Reference Hemisphere

The projection of the reference hemisphere is used to represent the spatial orientation of lines and/or planes and their angular relations to each other. Different types of such projections (also nets) are available, however, their use is not principally different. In each of these nets the orientation of a plane can be represented either by its pole (i.e. one point in the net) or by its great circle. The graphical details of such possibilities are given on Figs. 1 and 2. Conventional concepts of engineering mechanics, such as polygons of forces and shear strength along a plane surface, are combined with the basic graphic tool of the reference hemisphere to cover three-dimensional problems. The "friction cone" concept first proposed by Talobre represents the first of such combinations being utilized. It is briefly described and illustrated in Figs. 3 and 4.

Friction Cone by Talobre

The friction cone concept by J. Talobre is used in the approach described herein to represent or determine the angle between a line giving the direction of a force and a line normal to a plane. Combined with the angle of friciton along this plane, this approach can be used to evaluate graphically the possibility of sliding along this plane under a load acting in any direction.

Fig. 3 illustrates this concept, reduced to the plane case for clarity. Fig. 4 presents the same concept expanded to three dimensions. The projection of the traces of different cones on the equatorial net is shown in Fig. 4 (a). The example shows a great circle representing a plane p with its pole, P, with cone tracings about this pole for angles ϕ of $10°$, $20°$, and $30°$. The "cone surface" represents the directions of all lines which have a certain angle with the normal to the plane given by the pole. The intersections of an arbitrary plane through the pole, P, represented by its great circle, with a cone surface produces the directions of the two lines which are located within this arbitrary plane and have an angle, ϕ , to the normal of the plane p. Fig. 4 (b) shows an example for a "friction" angle of $20°$ in spatial view, with the directions of the two lines cited above labeled C and C'.

The tracings of a friction cone on the net is determined by plotting several arbitrary great circles through the pole and marking off the respective angles along them. The points of equal angles are then connected, to give the somewhat distorted (with the exception of a cone about the origin) projection of a circle in the equal area net.

Engineering Problem

The basic problem to be covered here is that of a rock wedge sliding on one or two geologic planes of weakness. It is essentially a slope stability problem, however, it also has been used in analyzing the stability of rock masses supporting arch dams, and to investigate the possibility of rock falls from walls of underground openings. The following aspects can be covered:

1.   Failure   pattern
1.1   Plane wedge formed by two plane sliding   surfaces
1.2   Modified wedge bounded by two mean en-echelon failure planes (see Fig. 11)
2.   Forces   acting   on   wedge
2.1   Gravity force
2.2   Seismic acceleration force
2.3   Hydrostatic forces acting on one or two bounding planes
2.4   Retaining forces introduced by rock tendons or rock bolts

3.   Shear strength along geologic plains each potential sliding plane
3.1   Friction only
3.2   Friction and technical cohesion

      Furtheron the hemispheric approach proved to be useful in the analysis
of
  a) Overtoppling (overturning) of subvertical columnar rock elements formed by
     the geologic structure, and
  b) Direction of borings in relation to the geologic planes for drainage, grouting,
     and rock tendons.

Basic Method of Analysis
      In the following pages the four basic steps of the proposed method of
analysis are presented using a simple sliding wedge problem: The steps are
  a) Digest of the geologic input data
  b) Establishing of geotechnical model and kinematics of sliding
  c) Establishing of active forces
  d) Establishing of equilibrium limits with consideration of passive forces.

      The approach in its most basic form is non-dimensional, assuming
only gravitational or gravity-related loadings, such as seismic acceleration, and
frictional resistances along the planes of weakness. However, further development of
the basic concept allows now to consider cohesion along potential slide planes and
specific external forces acting on the potentially unstable rock mass, such as dam
thrusts, hydrostatic thrusts, and supporting forces introduced by post-tensioned
rock tendons.
      The method of analysis consists of the following steps which are
illustrated in a slope stability problem.
  a) *Digest of the geologic input data,* Fig. 5 illustrates the problem posed and the
     pole diagram representing the orientations of the given planes of weakness,
     using an equatorial net in equal-area projection
  b) *Establishing of geotechnical model* with the determination of modes and
     directions of failures geometrically possible. Fig. 6 presents the concept for the
     above example in the hemispheric net relating the direction of active resultants
     with the thereby produced modes of failure. The spatial diagram serves to
     illustrate the general concept.

c) *Establishing of active forces* acting on a sliding wedge of given dimensions. Fig. 7 illustrates the use of supplemental polygons of forces to determine the active resultant due to weight and hydrostatic thrusts acting normal to the two planes of weakness. The same graphic procedure is used to consider any other active external forces.

d) *Establishing of equilibrium limits* utilizing the friction cone concept. Fig. 8 illustrates the friction cone concept originated by Talobre and the resulting stability limits. In the example solely frictional resistance is assumed for the first set of joints represented by pole P_1. Friction and cohesion is considered for the second set of joints, requiring supplemental planar polygons of forces. The stability factors can be assessed by inclusion in both friction cones and cohesive resistance, thus modifying the equilibrium limits.

Specific Aspects

After the basic method two specific aspects are presented

a) Modified failure wedge bounded by two en-echelon "mean failure planes" after Jennings. This concept, with its representation in the hemispheric projection is illustrated on Fig. 9, 10, and 11.

b) Graphical analysis of wedge sliding along the intersection of two "mean failure planes", with shear resistance consisting of both friction and cohesion. The sequence of analysis with hemispheric projection and supplemental planar polygons of forces is given on Fig. 12 and its comments.

Mean Failure

Jennings' concept of the mean failure surface can easily be introduced to the hemispheric representation. Jennings' definition of Fig. 9 lead to the representation of Fig. 10.

(1) The orientation of the α joints representing the shear plane is expressed by pole P_α. The shear parameters, the apparent friction angle ϕ_α, and the hypothetical cohesive strength k are referred to this pole, resulting in the cones of apparent friction and total shear resistance, respectively.

(2) The vector of the movement along the α-joints is normal to the pole P_α.

(3) The mean failure surface, β-plane, is used to determine both the weight of the potentially sliding rock mass, W_β, and the length or area effective for k, L_β or A_β, the latter for the three-dimensional case.

Failure Pattern

Fig. 11 represents the difference between the modified failure wedge formed by two mean failure planes, possibly complemented by deep tension cracks paralleling the ϕ-joints, with failure surfaces following the shear planes α_1 and α_2, It is believed that for large-scale failures the modified wedge is quite realistic. The plane wedges appear to be limited to small-scale stability problems.

Three-dimensional Analysis

Fig. 12A and 12B represent the comprehensive stability analysis for a simple example. It should be noted that only directions are represented in the reference hemisphere, whereas forces are determined in the supplemental polygons of forces which are based on angular relations determined in the hemisphere. The analytical procedure consists of the following steps.

(1) The poles $P_{\alpha,1}$ and $P_{\alpha,2}$ represent two sets of α-joints, α_1 and α_2 which are part of β-planes and along which shear is to take place.

(2) The direction of movements along the α_1 and α_2 planes is given by point $I_{\alpha,1.2}$ representing the intersection of the two α-planes.

(3) The first supplemental polygon of forces, see Fig. 12B, in the plane of section I, serves to divide the arbitrary loading, including weight, of the rock mass defined by the two β-planes, R_β into the components normal to the direction of movement, P , and the driving force, DF , in direction of $I_{\alpha,1,2}$.

(4) The second polygon of forces, section II, serves to determine the normal forces of both α-planes, $P_{\alpha,1}$ and $P_{\alpha,2}$.

(5) The polygons of forces sections III and IV, respectively, serve to assess the resisting forces developed along the two shear planes. The frictional component is given by $P_\alpha \tan \phi_a$, the cohesional component by A_β (k) , with A_β derived from the geometric configuration of the β-planes. The angles of total shear resistance, ϕ_Σ , can now be determined.

(6) Planes, i.e. great circles, through the points F_α and S_α, respectively, represent equilibrium limits; F_α for friction only, S_α for total shear resistance. These limits divide the 'safe' and 'unsafe' zones on the hemisphere.

(7) The factors of safety can now be determined by comparing the resisting forces along both α-planes, ΣRF , with the total driving force, DF . In the present example the factor of safety considering friction only is 1.0 (with the resultant R_β located on the respective limit line). A simple computation based on the results of the polygons of forces results in a factor of safety of the order of 2.0

for the total shear resistance.

(8) The factors of safety can also be graphically represented by adding modified equilibrium limits, using

$$\tan\phi_a' = \tan\phi_a/FS \quad \text{and}$$
$$k' = k/FS$$

The primed values, obtained by dividing the actual parameters by different factors of safety, are plotted on the hemisphere resulting in a family of limit lines with different factors of safety.

Overturning Problem in Slope Stability

Whereas the sliding problem, either planar or spatial, is often over-emphasized in engineering slope stability analyses, the stability probelm arising from overtoppling of steeply inclined columnar rock elements is often completely neglected.

This problem and the possibility to utilize hemispheric projections in its analysis are illustrated on Figs. 13 and 14.

Overturning Failure in Three Dimensions

Fig. 14 represents an analysis of the overturning problem in three dimensions. The reference hemisphere shows the three poles of an orthogonal joint system. More general systems can also be considered. The approximately rectangular pattern about pole P_1 represents the 'safe' range with respect of overturning. It is formed by the tracings of two overturning wedges on the surface of the reference hemisphere, with P_1 giving the direction of their common axis. Any forces with directions falling in this zone will not result in overturning. Section I shows a plane section through one of the overturning wedges with respect to planes p_1 and p_2 As can be seen, the wedges are defined by the centre angles δ_c with respect to the pole P_1.

The hemispherical plot of the chosen example shows the following:

(1) The weight of the rock element, which acts in a direction through the centre of the hemisphere, falls outside of the safe limit. Thus, it would produce overturning of the element about the northerly edge formed by planes p_1 and p_2 (see also plan view of actual rock element in lower left of Fig. 14).

(2) Any free surfaces, either natural or man-made (which are not shown on the

hemisphere for clarity) striking about EW and dipping towards N would be effected by this overturning mechanism.

(3) Because of the given geometry of the rock elements, overturning about the p_1 and p_3 edge is practically excluded.

The above conclusions indicate that a slope not subject to primary shear failure can represent a stability hazard, depending on the geometry of the surficial rock elements. In civil engineering applications such conditions are generally corrected by installation of rock bolts and/or post-tensioned rock tendons.

Conclusions

In several years of wide-spread use the geologic tool of the reference hemisphere has demonstrated its usefulness in three-dimensional stability analyses involving discontinuous rock system. It is now considered a routine method of analysis in engineering practice, particularly in professional environments in which geologists and engineers collaborate very closely.

It should be noted that the illustrated examples of the application of the reference hemisphere in engineering analyses do not at all limit the potential of such combination. Continuously new aspects are being introduced.

REFERENCES

[1] Talobre, J., "La Mécanique des Roches (Rock Mechanics)", Dunop, Paris, 1957, pp. 39-44.

[2] Terzaghi, K., "Stability of Steep Slopes on Hard Unweathered Rock," Geotechnique, Vol. 12, Dec., 1962, pp. 251-270.

[3] Müller, L., "Der Felsbau (Rock Construction)," Enke, Stuttgart, Germany, 1963, pp. 263-264.

[4] Goodman, R.E., "The Resolution of Stresses in Rock Using Stereographic Projection," International Journal of Rock Mechanics and Mining Sciences, Vol. 1, 1963, pp. 93-103.

[5] Friedman, M., "Petrofabrics Techniques for the Determination of Principal Stress Directions in Rocks," Proceedings, Conference on the State of Stress in the Earth's Crust, W.R. Judd, ed., American Elsevir Publishing Co., New York, 1964, pp. 451-550.

[6] Wittke, W., "Verfahren zur Berechnung der Standsicherheit belasteter und unbelasteter Felsböschungen", (In German), Felsmech. u. Ing. Geol., Suppl. II, Springer, Vienna, 1965, pp. 52-79.

[7] Goodman, R.E., and Taylor, R.L., "Methods of Analysis for Rock Slopes and Abutments: A Review of Recent Developments," Failure and Breakage of Rock, C. Fairhurst, ed., AIME, New York, 1967.

[8] John, K.W., "Graphical Stability Analysis of Slopes in Jointed Rock", J. Soil Mech. and Found. Div., Proc. ASCE Vol. 94, SM2, March 1968, pp. 497-526, with discussions, and closure, Vol, 95, SM6, Nov., 1969, pp. 1541-1546.

[9]   Londe, P., Vigier, G. and Vormeringer, R., "Stability of Rock Slopes, a Three-Dimensional Study", J. Soil Mech. and Found. Div., Proc. ASCE, Vol. 95, SM1, Jan., 1969, pp. 235-262.

[10]   John, K.W., "Engineering Analyses of Three-Dimensional Stability Problems Utilizing the Reference Hemisphere", Paper 7-16, 2nd Intern. Congr. Int. Soc. Rock Mech., Beograd, Sept. 1970.

[11]   McMahon, B.K., "A Statistical Method for the Design of Rock Slopes", Proc. 1st Australia-New Zeland Conf. Geomech., August 1971.

[12]   Londe, P., Vigier, G. and Vormeringer, R., "Stability of Slopes – Graphical Methods", J. Soil Mech. Found. Div., ASCE, Vol. 96, No. SM4, 1970.

[13]   Jennings, J.E., "A Mathematical Theory for the Calculation of the Stability of Slopes in Open-Cast Mines", Proc. Open Pit Mines Symp., Johannesburg 1970, Ed. P.W.J. van Rensburg, S. African Inst. Mining and Metall. 1970.

[14]   John, K.W., "Three-Dimensional Stability Analysis of Slopes in Jointed Rock", Proc. Open Pit Mines Symp., Johannesburg 1970, Ed. P.W.J. van Rensburg, S. African Inst. Mining and Metall. 1970.

[15]   Heuzé, F.E. and Goodman, R.E., "Three-Dimensional Approach for Design of Cuts in Jointed Rock", Proc., 13th Symposium on Rock Mechanics, Univ. of Illinois, Ed. E.J. Cording, ASCE, 1972.

[16]   Hoek, E., Bray, J.W. and Boyd, J.M., "The Stabilty of a Rock Slope Containing a Wedge Resting on Two Intersecting Discontinuities", Imperial College, Rock Mech. Res. Report, No. 17, April 1972.

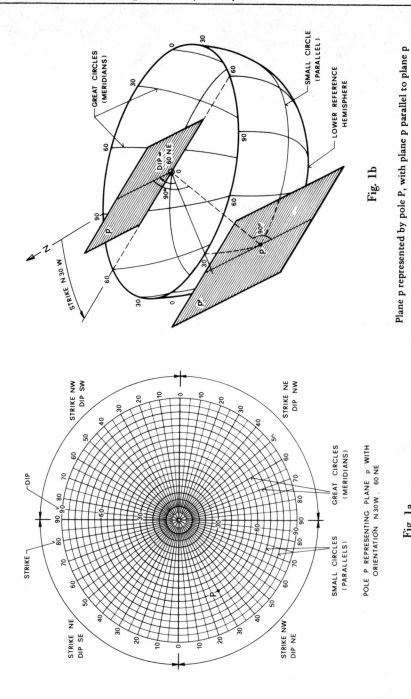

Fig. 1b

Plane p represented by pole P, with plane p parallel to plane p

Fig. 1a

Pole P representing plane p with orientation N 30 W, 60 NE

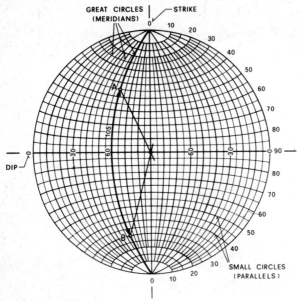

Plane upon which the points A and B are located dips 60 degrees.
Angle between A and B equals 105 degrees.

Fig. 2a

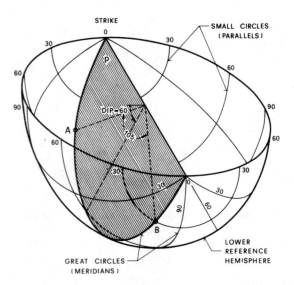

Fig. 2b

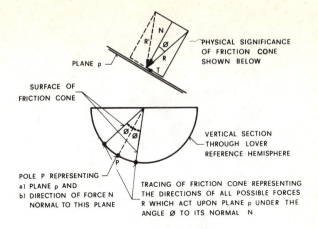

Fig. 3

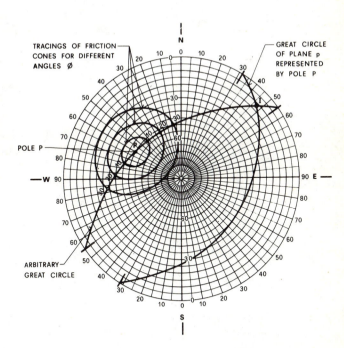

Fig. 4a

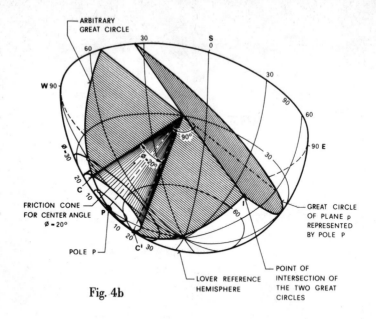

Fig. 4b

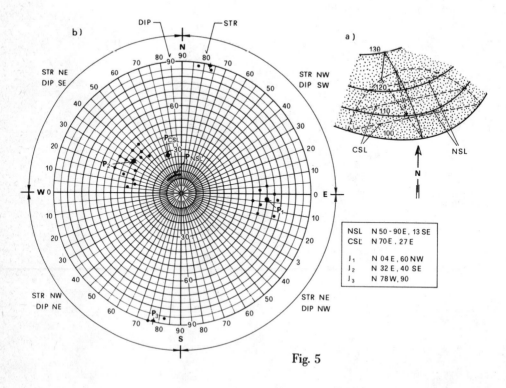

Fig. 5

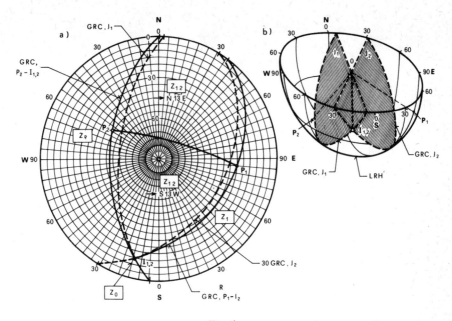

Fig. 6

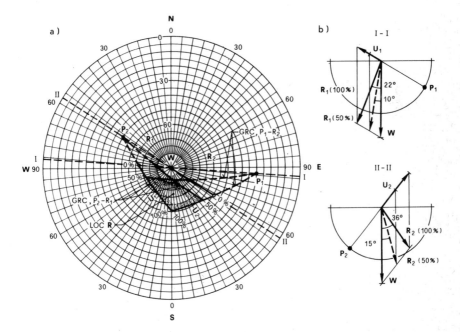

Fig. 7

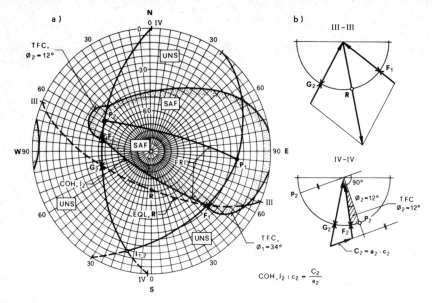

Fig. 8

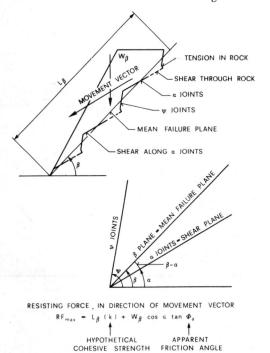

Fig. 9

Shear failure in jointed rock along mean plane after Jennings.

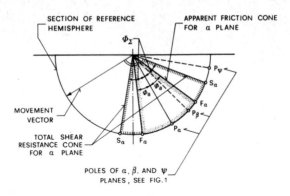

Fig. 10

Shear along mean plane represented in plane section of reference hemisphere.

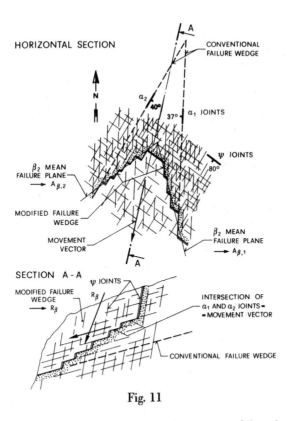

Fig. 11

Failure pattern of modified wedge bounded by two mean failure planes.

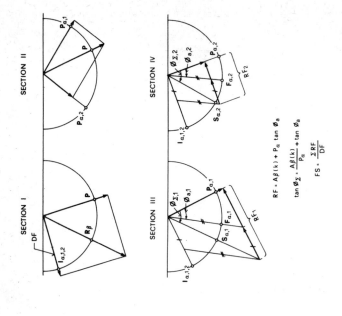

RF = A β (k) + P$_a$ tan $\emptyset_a$

tan $\emptyset_\Sigma$ = $\frac{A\beta(k)}{P_a}$ + tan $\emptyset_a$

FS = $\frac{\Sigma RF}{DF}$

Fig. 12 B

Supplemental planar polygons of force to Fig. 5a

Fig. 12 A

Equilibrium limits for wedge failing along combination of two mean
failure planes using reference hemisphere.

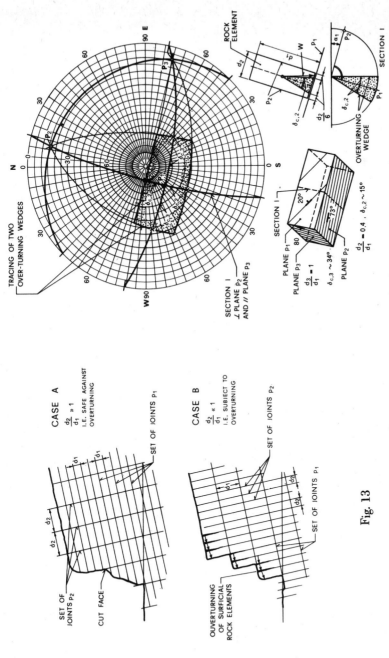

Fig. 14

Rock element subject to overturning using reference hemisphere.

Fig. 13

Joint spacing governing secondary failure of slope surface by overturning of rock elements.

ROCK STABILIZATION

P. EGGER
Abteilung Felsmechanic
Universität Karlsruhe

---

* All figures quoted in the text are at the end of the lecture.

I INTRODUCTION

From time to time a natural rock slope fails without any interference of man. Frequent are the cases where a mountain is just stable in its natural conditions, but is unable to support a supplementary load exerted e.g. by an arch dam. In a similar manner, the construction of underground cavities creates open surfaces in a rock mass which may be fissured and higly stressed being thus unable to withstand the new sollicitations. All these phenomena gave rise to the development of rock stabilisation techniques.

When speaking about stabilizing a rock, it should not be forgotten that the question is not only to make an unstable rock stable, but first of all keep a stable rock mass stable.

II PROPHYLACTIC MEASURES – CONSERVATION OF ROCK QUALITIES

In the lectures given by Mr. Rummel, Müller and John the mechanical properties of the rock element and the rock mass are treated, especially their stress-strain-laws exhibiting generally a rapid strength decrease in the Post-Failure--region (Fig. 1a). Furthermore the peak value for the strength obtained in a quick test is often higher than in a slow test (Fig. 1 b).

These two facts direct the attempts to conserve the qualities of the intact rock mass essentially to

1) avoid strains exceeding the strain at failure notably,
2) support the rock quickly enough after excavation

A) DESIGN STAGE

The prophylactic measures of rock stabilization being already in the stage of design of the works. The shape of the rock surfaces, independently whether they are underground (Fig. 2) or in an open pit (Fig. 3) should always be chosen so as to avoid too high **stress concentrations** — what means local high strains and loss of the rock strength near to the surface, phenomena accompanied in brittle rocks by spalling and slabbing. The excavation should, therefore, be rounded and adequate to the surrounding stress field.

Because a fissured rock is practically unable to resist to **tensile stresses**, the shape of an excavation in rock should be designed accordingly. Otherwise there will be zones of loose rock, unstable during excavation and acting as loads on to the support (Fig. 2).

B) CONSTRUCTION STAGES

In the attempt to conserve the initial rock qualities the influence exerted by the construction parameters is often underestimated.

1) Sequence of excavation

In many cases, an underground work cannot be excavated in full section but has to be done by partial sections. It is evident that the execution of each partial section creates stress redistributions and certain displacements in the rock. These are followed by local deteriorations of the rock mass, essentially near the corners of the partial excavation. Difficulties during the excavation of subsequent stages and a general decrease of the stability of the underground opening may be the consequences of an improper choice of the excavation sequence (Fig. 4). On the other hand by a careful layout of the different stages (Fig. 5), these negative consequences can be minimized.

2) Methods of excavation

In rock engineering, great attention has to be paid to the method of excavation, not only from the economical standpoint but also with regard to the possible alterations of the rock quality. The most generally practicable method is blasting, and the experience shows that the use of too heavy charges, the choice of exaggerated lenghts of unit advancement and an unfavourable arrangement of the boreholes frequently lead to a serious demolition of the rock near the new open surface and to heavy overbreak. On the other hand, specially developed procedures such as "Smooth blasting" or similar limit the impact of the explosion to a reasonable value.

If the rock is not too heterogeneous and not too hard or abrasive, the use of tunnelling machines has proven very successful in the recent years. This mechanical excavation without any shock-waves is best suitable to conserve the rock qualities and thus to minimize the need of rock support. There is one remark, however, to be indicated namely that in rock which tends to overbreak even in case of machine tunnelling, the unstable rock blocks frequently fall down without warning, behind the machine (Fig. 6). When using explosive, these blocks come already down during the blasting.

3) ROCK SUPPORT

A suitable rock support built in quickly enough after the excavation is in many cases rather a prophylactic measure against the deterioration of the rock than a support against active rock pressure. In order to enable the support to prevent rock disintegration subsequent to the excavation it has to satisfy to the following conditions:

— A quick effect is needed
— Excessive local strains must be prevented.

These conditions are well fit by a support consisting from shotcrete, rock bolts and steel ribs applied separately or in a combined way as required.

In case of water inflows, this water has to be drained quickly in order to prevent retrogressive erosion in the water bearing fissures or the softening of sensitive rock. In the latter case it is always recommended to apply rapidly a thin layer of gunite onto the rock surface for sealing it against the action of atmosphere.

Shotcrete: is a mix of aggregates up to about 15 mm (Fig. 7), cement and an accelerator, the water being added either in the mixer or at the end of the flexible tubing. It is important to obtain a mix with a relatively high strength already after several hours (Fig. 8), in order to create quickly an efficient support which is in intimate contact with the rock. Contrarily to the classical steel support with wooden lagging, the shotcrete provides a continuous support (Fig. 9) preventing also local free rock displacements which would lead to disintegration and loosening.

Rock bolts: The action of the usual rock bolts (steel bars of about 2 to 6 m length) is similar to that of the shotcrete: they hold more or less unstable rock blocks in place, prevent them from moving along the fissures of from falling out, keeping thus the rock fabric intact and preserving its strength (Fig. 10).

Steel ribs: together with a wire-mesh, they serve as a first protection of the miners. After the projection of the shotcrete they act mainly as reinforcement in case of bending moments due to local loads and as an additional resistance against shear.

This brief review of measures to conserve as well as possible the original rock qualities did not intend to give a detailed prescription how to proceed in an individual case, but was given with the aim of drawing the attention on the fact that by careful design and construction much stabilization work — in the common, i.e. therapeutic sense — can be avoided.

III THERAPEUTIC MEASURES – IMPROVEMENT OF THE GIVEN SITUATION

Despite all the above mentioned precautions, it often occurs in rock engineering that stability of the construction cannot be obtained without measures which improve the given situation.

There are two main groups: The first one comprises the measures which improve the rock qualities – essentially grouting (Fig. 11), in special cases freezing (Fig. 12). The techniques of the second group tend to modify the **stress field**. Prestressed anchors (Fig. 13) and a systematic drainage (Fig. 14) belong to it.

A) GROUTING

1) INTRODUCTION

Rock grouting though used since several decennies has still remained an art rather than an applied science. Its aim is double: either it shall confer to the fissured rock a higher strength (consolidation) or a lower permeability (impermeabilization), or both of them.

The following brief outline refers essentially to the work of Prof. H. Cambefort.

2) REQUIREMENTS
a) Mechanical parameters of the grout

As rock grouting consists essentially in filling fissures – under low or high pressure – the grout pumped into the fissures has to fulfil certain mechanical requirements in order to be suitable to consolidation or tightening.

aa) Compression

What strength must a grout have in order not to be squeezed out by a compression stress? (Fig. 15) Prandtl found for a perfectly plastic medium with a cohesion c in a fissure of length 1 and width e the allowable average compression stress p:

$$p = c \left(\frac{\pi}{2} + \frac{1}{2e} \right) \tag{1}$$

As 1/e is always large, even very small cohesion values are sufficient for the grout to resist to high compression stresses.

ab) Hydraulic pressure

   Similarly the resistance against being pushed out by a water head (Fig. 16) was found by Mandel to be

$$q = 2C \frac{1}{e} \quad .....(2) \text{ for a purely cohesive grout and}$$

$$q = \frac{C}{tg\rho} \left(e^{K \cdot \frac{1}{e}} - 1 \right) ......(3) \text{ for a medium with cohesion } c \text{ and internal friction } \rho$$

with $\quad K = 2\ tg^2\rho \left(cot\rho + \rho + \frac{\pi}{2} \right) \quad ..... (4)$

   In a fissure with a large value of 1/e, even a very poor grout is able to resist to high hydraulic pressures.

ac) Shear

   Whereas no problem exists generally for the grout to resist to compression and to a water head, a smooth grout filled fissure which is sollicited to shear yields at

$$\tau = C + p \cdot tg\rho \quad ..... (5)$$

where c and ρ are either the values of the grout itself or of the contact plane between grout and rock (Fig. 17). It is evident that a consolidation grouting which has to confer to the rock mass a high shear strength must be done with resistant grout, generally based on cement. This is however not absolutely necessary if the fissures are rough and well indented so that the shear strength is furnished by the rock (Fig. 18).

ad) Washing out

   A plastic grout even with a small cohesion was seen to resist very well to hydraulic pressures. This is not the case, however, for grout types which are viscous liquids such as bitumen. Here, the hydraulic pressure creates a slow but steady flow and finally pushes the grout out of the fissure.

b) Economy of grouting

Besides the mechanical requirements, a grout should be of easy confection, well pumpable and, last not least, inexpensive.

Depending on the aim of grouting and on the width of fissures, a variety of grout types have been developed. For reasons of economy, cement grouts or grouts on a cement basis are the most commonly used mixes, giving satisfaction in the majority of cases encountered in rock grouting.

3) GROUT TYPES

There are three different kinds of grout types:
— unstable mixes,
— "stable" mixes and
— liquids.

a) Unstable grout mixes

The unstable mixes are essentially suspensions of cement in water. Sometimes, rock powder is added for economic reasons. They are homogeneous as long as they are stirred. At the end of the movement sedimentation begins. Similarly when the grout is transported in pipes of large diameter, i.e. with low velocity, sedimentation occurs obstructing the pipe.

b) "Stable" mixes

The so-called "stable" mixes are also essentially suspensions of cement or clay in water, they remain homogeneous after the grout confection for several hours, i.e. at least for the duration of the grouting. In many cases, these mixes are thixotropic and rigidify when the movement comes to rest. Their aspect is that of a viscous, slightly rigid liquid. Generally the setting time is — contrarily to the unstable mixes — superior to one day, i.e. much longer than the usual grouting process.

For temporary works, also specially treated emulsions of asphalt in water are used. A careful choice of an emulgator and a coagulator allow this grout to penetrate into the rock fissures and to rigidify by breaking of the emulsion, after a controlled time.

c) "Liquid" grout types

For sake of completeness, it shall be mentioned that a grout is commonly called "liquid" if it contains particles of a hardly measurable size. For instance, silica gels and organic resins belong to this group. These grout types have a very low viscosity down to 2 cP during the grouting time and are able to penetrate nearly everywhere where water is circulating. Besides, the setting time can be regulated very well (Fig. 19). Because of the relatively high costs of these grout types, their application in rock grouting is limited to special cases, e.g. for the impregnation of porous rock or for grouting rock bolts which must be tensioned very quickly.

4) GROUT PROPAGATION IN FISSURES

After this brief summary of frequently used grout types, the main phenomena occurring during the grouting process shall be examined:

a) Flow of "stable" and liquid grout types

Contrarily to the grouting of loose soil where a volume element of grout flows through voids of largely varying sections, in rock grouting the mix flows essentially through fissures (Fig. 20). If the fissure is assumed to be of constant width e, the following approximative relation for the laminar flow of a Newtonian fluid can be derived:

$$p_o - p = \frac{6\,\nu\,Q\,\gamma}{\pi\,e^3}\,\ln\frac{r}{r_o} + \frac{3\,\gamma\,Q^2}{20\,g\,\pi^2\,e^2}\left(\frac{1}{r_o^2} - \frac{1}{r^2}\right) \text{(BAKER)} \ldots \ldots (6)$$

with p_o ..... pressure in the borehole of radius r_o
      p ..... pressure at a distance r
      ν .....coefficient of viscosity
      Q .....flow rate
      γ .....density of the mix

The first term allows for the pressure loss due to friction, the second one due to the modification of kinetic energy in radial flow.

This relation shows that with stable or liquid grout types there will always be a flow as long as the viscosity does not tend to infinity.

Assuming that at a distance r' (= radius of action) the pressure p = 0, we obtain for a given fissure and grout type a diagram of the type given in Fig. 21, which has to be slightly modified for greater values of Q when the laminar flow changes into turbulent flow. Anyway the influence of the borehole diameter, but first of all that of the width of the fissure can clearly be seen.

b) Widening of fissures

The foregoing derivation was based on the assumption that the width of the fissure is constant. When we now take the elastic half-spaces be pressurized, Boussinesq's formula for the settlement w of a plate of radius r under the load p gives us a rough idea of the widening of the fissure:

$$w \cong 1,5 \; \frac{p \, r}{E} \quad \ldots \ldots \; (7)$$

On he other hand, the integration of equation (6) between the borehole and the radius of action yields the total force and, thus, the average pressure $\bar{p}$ acting on the fissure. Its increase in width can now be evaluated by inserting $\bar{p}$ into eq.(7). Despite this rough approximation, it is clear that now the relation between the pressure in the borehole and the flow rate is considerably different from the case with constant width e of the fissure (Fig. 21).

This phenomenon of widening the fissure by the hydraulic pressure makes the quantitative interpretation of the water tests very difficult. But it is the same phenomenon which enables to grout a fissured rock efficiently with unstable mixes.

c) Flow of unstable grout

The flow of an unstable grout mix is a hydraulic transport of solid particles. Investigations showed that particles of the size of cement grains begin to sediment at a flow velocity of 3 to 4 cm/s. However velocities of 20 to 30 cm/s are necessary to bring these particles in suspension again.

When a fissure in rock is grouted with an unstable mix, different stages can be observed (Fig. 22):

a) In the first stage, the flow through the fissure follows approximately the eq. (6). Pressure and mean velocity decrease monotonously with the distance from the borehole. In a cross section, the velocity is maximum along the centerline and zero along the walls .

b) The cement grains near the walls tend to fall out from the suspension. Those which are near the upper wall fall down into a zone of velocity which is high enough to carry them away. The grains in the lower part of the fissure however can sediment and, thus decrease the section. The velocity increases at that place, but is not sufficient to erode the sedimented grains.

c) Behind these sediments, the velocity decreases abruptly and gives rise to a new sedimentation downstream of the existing one.

d) The longer the cement deposits become, the higher the pressure increases in front of them. From a certain value on, the flow rate of the grouting pump diminishes, the grout velocity decreases and a sedimentation of cement occurs upstream of the original deposit until reaching the borehole. Then the grouting process is completed.

It should be noted, however, that there remains still a narrow slot where a small grout quantity can flow. The hydraulic fill is not perfect. On the other hand the original fissure was widened by the grouting pressure and will tend to contract after the end of grouting. If the grouting is performed with a high pressure, e.q. 50 to 100 bars, the widening of the fissure is considerably greater than the remaining slot width, and the rock will be completely sealed and even prestressed by the grouting.

5. CRITERIA FOR THE CHOICE OF GROUT TYPE

Seeing the variety of grout types which are available to the engineer, there is a need of criteria leading the choice of the grout type appropriate in a specific case.

These criteria are necessarily based upon the knowledge of the nature of the rock: distance, orientation, width and spacing of the fissures, the existence of big voids (cavities) or of crushed rock zones and, last not least, the permeability of the rock mass are the most interesting parameters.

a) Borehole inspection and water tests

These informations can be provided by the inspection of cores, by the optical (or TV) inspection of boreholes and by water tests. The advantage of the optical inspection (as well as the core drilling method after Rocha — Fig. 23) lies in seeing the width of the fissures in place and in showing whether a core loss was due to a void or to a weak filling material (gauge) washed out during drilling. The water tests or Lugeon tests (Fig. 24) are performed by sealing a portion of the borehole of several meters length by one or two packers and by recording the flow rates of water pumped into the borehole at different pressures (Fig. 25). The type of diagram obtained gives certain indications about the nature of the fissures; the Lugeon-value, i.e. the flow rate per minute at $p = 10$ kp/cm^2 for $L = 1$ m borehole length, is a frequently applied criterion for the necessity of grouting.

The choice of the grout type depends essentially on the width of the fissures. The Lugeon test is however not able to furnish them, as a certain flow rate can be obtained when strucking e.g. one large fissure or a series of narrow ones. Therefore the Lugeon-value of a rock cannot be considered as an absolute criterion for indicating grout type and consumption, it must be combined with grouting tests.

b) Grouting tests

The only way of finding out with certainty the suitability of different grout types is to perform in-situ-grouting tests. Tests blocks (Fig. 26) are grouted with different grout types, the quantities of taken grout and the pressures are recorded; at the end, control drill holes are executed in order to examine the cores and to perform comparative water tests.

c) Scope of grouting

Besides the criteria given by the groutability of the rock, the scope of grouting (consolidation or impermeabilization) influences also the choice of grout type. A consolidation grout will always be chosen so as to obtain a certain strength, whereas the strength may be low in a grout screen serving only as a tightening.

6) GROUTING TECHNIQUES
a) Grouting with unstable mixes

In rock grouting the use of unstable cement mixes is most frequent because of the fact that these mixes are very economic, easy in the confection and well groutable into fissures which are not too wide. They are appropriate as well for consolidation (pure or preponderantly cement mix) as for sealing (addition of clay, ashes etc.).

aa) Definition of maximum grouting pressure

At the begin of the grouting work, the maximum pressure which shall be reached, has to be defined. This question gives frequently rise to arguments, and different traditions have developed. In the anglo-saxon countries, tendency goes to rather low pressures, e.g. 1 p.s. i/ft = 0,24 kp/cm^2 m depth, whereas in Europe generally much higher pressures are used, e.g. 1 kp/cm^2 m depth. The reasons for these contrary tendencies can be seen in the facts that on the one hand the fear to create upheavals of the surface and to waste grout is predominant, on the other hand the knowledge about the efficiency of grouting.

When grouting fissures in small depth which are parallel to the surface (Fig. 27) care must be taken at the condition not to grout a series of adjacent boreholes simultaneously but to grout one hole after the next one. In this case, the volume of rock which would be moved by the grouting is generally much larger than just the overlying cylinder. Only in special cases, when the ration between the radius of action R' and the depth D is large, e.g. in the presence of karst phenomena near the surface, a limitation of the pressure to about the overburden pressure is justified. Anyway, when grouting near to existing constructions, an observation system to detect movements should be installed permitting a continuous control of the grouting works.

When grouting fissures situated in greater depth, no upheavals of the surface need to be feared. **Consolidation** grouting should be done with high pressures in order to widen the fissures during grouting more than the width of the remaining slot and to create thus a prestressing effect in the rock. High pressure is therefore the best warrant to obtain a good result.

For **impermeabilization** grouting, the prestressing effect is of minor importance, as we know that anyhow a perfect impermeabilization is impossible. Therefore, it is reasonable to limit the maximum pressure to several bars above the pressure of the water head acting in the most unfavourable conditions.

ab) Water-cement-ratio of the mix

After having defined the maximum grouting pressure, e.g. 50 kp/cm^2, the question is to impregnate the fissures in an optimal way, i.e. with a minimum cost. The first idea is to begin grouting with a thick mix in order to block the large fissures, and then go on with thinner mixes for sealing the small fissures.

Thus, the quantity of cement is minimized. The disadvantage of this method is that the borehole will be blocked as many times as there are different grout mixes, and must be redrilled.

In order to avoid this, the inverse way of proceeding is preferred. The results of water tests give a first rough idea about the W − C − ratio to be begun with without the risk of blocking the borehole. For example, with Lugeon-values inferior to 5 units, a W/C-factor of 8/1 generally gives satisfaction; for L.U. = 5 to 10, W/C = 4/1 is generally convenient. The usual technique consists in grouting the first, thin mix for about 15 to 30 minutes and to look if the pressure is rising. In the affirmative case, the grouting is continued with the same mix until obtaining the maximum pressure. In the negative case, a thicker mix will be used and again pumped for 15 to 30 minutes, and so on, until finally reaching the maximum pressure.

This procedure leads to a slightly higher grout consumption than the former one, but avoids the expensive and time consuming redrilling of the boreholes.

b) Grouting with "stable" mixes

When the water-tests show a high permeability, e.g. > 10 Lugeon-units, grouting will reasonably be begun with a "stable" mix which can be injected like a viscous liquid.

ba) Limitation of quantities

As no notable sedimentation occurs and the setting time is generally much longer than the grouting time, the groutable quantities are not limited by the grouting process, but are proportional to the pressure and inversely proportional to the viscosity. This has two consequences: first, the necessary grout quantity must be evaluated by geometrical considerations (number and width of fissures, radius of the zones to be grouted). Secondly, it is possible to grout at very low pressures when reducing the flow rate; this is precious for grouting near a surface.

bb) Necessary provisions

In order to obtain a full sucess of grouting with "stable" mixes, it must be remembered that these grout types are not perfectly stable; a certain sedimentation occurs before setting. As, furthermore, the fissures in the rock are practically not widened up when grouting at low pressures, the obturation of the fissures will not be perfect. To overcome this handicap two ways are possible: Either, the grouting is completed, after the setting of the "stable" mix, by a thin unstable mix; this needs redrilling of the boreholes. Or, a swelling agent is added to the mix which counterbalances the loss of volume due to sedimentation (e.g. Prepakt-mix); thus a treatment in a single stage is possible. The fine fissures are, however, not treated in this case.

c) Grouting with liquids
ca) Fields of application

For sake of completeness, grouting with liquids should be mentioned, although their principal fields of application are in grouting of fine grained soil. In rocks, only few special cases justify the relatively high costs of these grout types. One possible application is the crossing, by a tunnel in rock of a large fault filled with badly crushed material, in the presence of groundwater. Another special case is the quick realization of a bond between the rock and a rock bolt, by organic resins.

cb) Limitation of quantities

The viscosity of the liquid grout types being particularly low, the grout quantities must be limited and the setting time adjusted accordingly to the duration of grouting. The latter is important especially in the presence of moving groundwater in the fissures, in order to prevent the grout to be washed out before having set.

d) Combined grouting procedures

Under certain circumstances, it is necessary or at least preferable to use different grout types successively. Two cases occur quite frequently:

da) Treatment of very wide fissures

In the first stage, a bulkhead is created by a thick "stable" mix with — preferentially — thixotropic properties (Fig. 28); in the second stage, a thin unstable mix is grouted into the interior space limited by the bulkhead. Thus uncontrollable losses of grout are efficiently prevented.

db) Fissures in porous rock

In a fissured rock which itself is porous, i.e. permeable, grouting hits against certain obstacles: part of the water of a "stable" or unstable mix is adsorbed by the permeable walls of the fissures, the mix thickens and the flow is soon blocked. An efficient remedy against this phenomenon consists in grouting first a silica-gel which seals the pores of the rock adjacent to the fissures. Then, grouting is continued under pressure with an ordinary cement mix which pushes away the gel in the fissures. In the small voids of the rock, however, the gel will remain, sealing them efficiently, so that no water from the cement mix will be lost there.

7. EXECUTION OF GROUTING WORKS

In the following, a brief general outline is given concerning the sequence of operations involved with rock grouting.

a) Investigations

When for a planned construction in fissured rock problems arise concerning the strength or the permeability of the rock mass, the dimensions of the rock zone to be investigated have to be defined. They depend essentially on the nature of the rock and on the kind of construction.

If a **consolidation** of the rock reveals to become necessary by static reasons, the dimensions of the critical zone can roughly be evaluated by considerations about the stress field near the construction.

In the case of an **impermeabilization**, its required degree is an additional factor for the definition of the volume of investigations. E.g., as a thumb rule the depth of investigations to be carried out beneath a dam is generally said to be of the order of magnitude of the height of the dam.

As it was already pointed out, the investigations comprise essentially the study of geological maps, core drilling, open pits, exploratory shafts and galleries.

b) In-situ-tests

The exploratory works are followed by laboratory and in-situ tests. In this context, the Lugeon or water tests are of highest importance, as they are particularly appropriate to show quickly the most permeable zones.

In-situ grouting tests should be carried out in an early stage in order to direct the design into the right way.

c) Design

Based on the results of the exploratory works and of the tests, the design of the grouting campaign will define the extent and the nature of works. Criteria taken from the experience are of a great help for the practical design: e.g. in the case of a grout curtain beneath a dam, the necessity for grouting is generally seen where the water tests yield a Lugeon-value superior to 1 (l/m. min. 10 atm.). Thus, the depth of the boreholes is defined. (Another example: most types of prestressed rock anchors should not be built in at a Lugeon-value superior to about 2, otherwise the rock has to be improved by grouting).

When the radius of action of grouting a single hole is known the layout of the boreholes (distance, number of lines) can be worked out. By the interpretation of Lugeon and grouting tests, the appropriate type of grout mix which depends also to a great extent on the scope of the grouting, is indicated.

d) Grouting works
da) Drilling pattern

As the best campaign of investigations never can give detailed indications about the local conditions all over the site, the drilling and grouting pattern will be chosen so as to do, simultaneously with the grouting work, a continuous control of its efficiency. Frequently in a grout curtain every fourth borehole is executed and grouted in a first step (Fig. 29). In the second phase, the water tests performed in the boreholes on half way show whether or not the grout circulated until there. After grouting the holes of the second stage, the same procedure is repeated for those of the third stage.

This system permits to be flexible in the execution allowing to drop boreholes in the case the radius of action shows to be larger than assumed.

db) Ways of grouting a hole
α) Grouting the whole length

When the borehole is drilled, the question is how to grout it. The simplest way consists in grouting it at once over the whole length by sealing a tube to the borehole (Fig. 30). The inconvenients of this method are, however, that grouting cannot be done with high pressure due to the vicinity of the surface and that the risk of blocking the borehole is high due to the sedimentation of the grout in the borehole, when its length exceeds about 10 m.

Long boreholes are, therefore, better grouted by sections of several meters length.

β) Climbing sections

The sequence of the grouting sections may be twofold: Either climbing from the end of the borehole to its mouth or inversely (Fig. 31, 32). Grouting by climbing sections has the great advantage that the borehole is drilled over its whole length in one step, and that the drilling and the grouting shift are working independently from each other. This is the reason why this system is generally used wherever possible. It must be noted, however, that it cannot be applied when the rock quality is so bad that the packer does not hold; in this case, the pressure cannot be increased sufficiently and grout flow around the packer risks to occur.

γ) Descending sections

Grouting by descending sections does not show these inconvenients because the packer is always installed in the previously grouted and, thus improved section.

Furthermore, it allows to grout the first section near the surface with low pressure and special care, creating a consolidated roof for the underlying sections.

On the other hand, the borehole is blocked at each section and must be redrilled when going ahead to the next one. Drilling and grouting operations are therefore inseparable.

e) Controls

Due to the heterogeneity of the rock masses generally encountered, even the most experienced grouting specialist will carefully control the results of his work.

ea) Local

By the arrangement of the drill pattern such as described in the previous article, there is a continuous control given by the sequence of drilling and grouting. In addition to this, extra control holes are drilled at irregularly distributed places in order to take out rock cores and to perform water tests. All these controls are, however, local ones, comparable to needle stitches into the rock mass.

eb) Global

In order to know the overall behaviour, controls of larger dimensions are needed, but are unfortunately lengthy and expensive. The **consolidation** effect can be controlled by large scale bearing tests (plate or radial jack tests) or, in an indirect way, by seismic or ultrasonic tests.

The reduction of **permeability** of a fissured rock is difficult to control at a large scale, when it is question of a grout curtain beneath a dam. In the case of underground openings, the task is easier, the control may be obtained by driving a test gallery and observing the water inflow.

8. EXAMPLE
a) Situation

A horizontal gallery of about 6 m diameter, situated in 110 m depth in sedimentary rocks. The ground water table is 80 m above the gallery. Different sets of joints are encountered, one of them is perpendicular to the tunnel axis. All these joints are fairly closed and impermeable over a considerable tunnel length. From a certain point on, the vertical joints become wider, reaching a maximum width of several centimeters. As they are open (no filling was encountered), groundwater comes out from pilot holes drilled in the tunnel face at a quite important flow rate when the holes strike such open joints. After sealing the boreholes at their mouth full water pressure of 8 atmospheres builds up immediately.

b) Scope

The scope is to seal the more or less wide vertical fissures in order to proceed with tunnelling under acceptable working conditions.

c) System

A two stage system is chosen (Fig. 33):

a) Production of a cofferdam by grouting a limited quantity of a stable cement-sand-bentonite mix with a low cement/sand ratio for economic reasons. Addition of an accelerator in order to obtain a setting time of about 4 hours.

b) Completion of grouting with a cement mix with addition of some bentonite for obtaining a better pumpability.

The layout of the boreholes can be seen on Fig. 33.

B) FREEZING

1) INTRODUCTION

In water bearing ground, an improvement of the stability can also be obtained by freezing. The pore water becoming ice confers to the material a temporary cohesion. The application of this method in rock engineering has not been widespread due to the competition of grouting.

a) Types of ground

Freezing in rock engineering has been restricted to completely crushed zones filled with a quasi impermeable gauge not penetrable by grouting. This method was applied several times for the construction of mine shafts, but very seldom for tunnelling in rock.

b) Types of construction

Contrarily to grouting, rock improvement by freezing has only a temporary character. The tunnel lining has to be dimensioned to resist the full rock and water pressure corresponding to the situation without freezing.

2) METHODS
a) Principle of freezing

The principle of freezing constists in creating frozen ground zones around pipes entered into boreholes. In these pipes a cooling agent circulates (generally in a closed loop, Fig. 34) carrying away calories from the ground. The desired effect of the method is obtained when the frozen zones around the different pipes are touching each other without exception.

b) Types of cooling agent
ba) Brine

The most frequently used cooling agent is a brine of calciumchloride having a freezing point of about -37 ° C. Its specific heath is comparable to that of water.

bb) Nitrogen

Recently, several applications have been known where liquid nitrogen was used as cooling agent. Here, the freezing effect is much faster due to the low temperature of about -200°C despite the lower specific heath of liquid nitrogen. In this case, the nitrogen coming back from the freezing pipe is not cooled again and reused, but it escapes to the air.

c) Execution

The layout of the boreholes on a freezing site is comparable to grouting. Recent developments tend to perform subhorizontal holes by a method containing a steering head, avoiding thus the most dangerous borehole deviations.

However, the freezing method needs about several months time to build up the frozen ground zone. This requirement restricts the application of the freezing method to sites where enough time is available.

In the presence of flowing groundwater special investigations have to be undertaken in order to know its implications for the additional costs due to the increased heath transportation and for the modifications in the layout of the boreholes.

3) EXAMPLE
a) Situation
Gallery crossing a zone of crushed dolomite.

b) Reference
O. Rescher: "Die Anwendung des Gefrierverfahrens beim Ausbau eines Stollens in einer schwierigen Gebirgsstrecke". Rock Mechanics, Suppl, I, 97 -123 (1970).

C) ROCK ANCHORS

1) INTRODUCTION
Whereas the grouting and freezing techniques try to obtain rock stabilization by an improvement of the rock qualities (strength, permeability), a second group of therapeutic methods comprising the use of rock tendons and a systematic drainage are characterized by modifying the stress field.

a) Definition
Anchors are structural elements serving to transmit tensile forces into the rock; the anchors are inserted into a previously executed borehole, and in a subsequent stage, the bond with the ground is performed. Anchors do not transmit compressive forces.

b) Action of anchors
The action of anchors is twofold:

ba) Modification of stress field
As soon as an anchor is tensioned, the original stress field is modified (Fig. 35).

bb) Prevention of disintegration
At the same time, a tensioned anchor holding a rock block in its original position acts as a preventive measure against the disintegration of the rock (Fig. 36).

c) Fields of application
ca) External forces
The transmission of external forces to the rock at a certain depth (Fig. 37, 38) is an important field of application of rock anchors, though rather a measure to stabilize structures on the rock than the rock itself.

cb) Internal forces
The provisions with stabilizing forces of a rock zone which is not in equilibrium by itself, thus the stabilization of rock properly speaking, is the second field of application of rock anchors (Fig. 35, 39).

2) GEOMECHANICAL AND STATIC ASPECTS

a) Evaluation of anchor forces and direction
The ᵢ evaluation of anchor forces is simple as long as it can be done by purely static considerations without additional requirements due to kinematics (Fig. 35, 40). It is sufficient to undertake a vectorial equilibrium analysis.

In all cases where kinematic restraints must be allowed for (Fig. 36, 39) the evaluation of the necessary anchor forces becomes essentially more difficult. The dilation characteristics of the rock and the resulting change of the anchor forces have to be taken into account. In cases such as Fig. 36 this is comparatively easy, but for underground openings the design of rock anchoring is still rather tentative.

Concerning the optimal anchor direction, the variations of the necessary anchor length and force with the direction must be considered (Fig. 41). The variety of possible anchor directions is however strongly limited by the requirement that the resulting force should not deviate than by the angle of internal friction (Talobre's cone) from the normal vectors to the joint planes (Fig. 42). Commitments due to acceptable working conditions are also narrowing the spectrum of possible anchor directions. Too acute angles between joints and anchors may lead to important deviations during drilling.

b) Stability of the anchor — Bond length to ground
Tests were performed where the annular space between the tendon and the wall of the borehole was filled with pure sand. When pulling the tendon, there built up a notable resistance after some displacement (Fig. 43), phenomenon explainable by arching. The stresses transmitted to the rock surface are inclined to

the axis of the anchor, tending thus to burst the rock.

Example: Borehole diameter = 10 cm
           Anchor force A = 30 t/m → Shear stress ≅ 100 t/m²
           Assumption: Inclination = 45°→ Radial stress ≅100 t/m²
           Induced circumferential tensile stress ≅ -100 t/m²
           If there is not a higher compression stress due to the primary stress
           field, fissures in the rock will open.
           Rule: Grout the anchor with at least the same pressure, in order to
           prestress the rock before the anchor is tensioned.
           For a grouted anchor, there are two possibilities of failure:

ba) Bond between grout and rock

A rough but simple evaluation of the transmissible force consists in pushing a rock core through a mortar bed (Fig. 44). These tests yield values for the average shear stress along the core in the order of magnitude of 5 to 40 kp/cm² depending on the tested rock type.

bb) Bond between grout and tendon

This problem is well known from the theory of reinforced concrete. Only when using thick bars or tubes, this criterion may be less favourable than the former one.

bc) Necessary bond lenght

The design of the necessary bond length is rendered difficult by the fact that the distribution of shear stresses is not uniform along the bond length (Fig. 45).

Furthermore, differences exist between the anchor types where the tendon transmits the forces directly to the grout and those where a compressed structural member (Fig. 45, 46) does this transmission.

C) GLOBAL STABILITY – FREE ANCHOR LENGTH

The free anchor length is defined by the condition that the global stability of the system rock + anchor (+ support) is sufficiently high.

In the case of a rock slope, e.g. (Fig. 35) the safety against sliding along a potential combined sliding surface (A-D-E), passing behind the anchor ends, has to be examined.

When bolting the roof of a wide cavity (Fig. 47) in a jointed rock, then a simple and reasonable approach consists in considering the roof as a beam of height h, loaded by a loosening pressure p. The role of the anchors is to tighten the strata together in order to act as a whole. The height is to be calculated from the condition that no tensile stress occurs in the beam:

$$\sigma_A = \sigma_h - \frac{6\,M}{h^2} > 0 \rightarrow h > \sqrt{\frac{6\,M}{\sigma_h}}$$

with M = max, moment due to external load p

In more general cases, however, the foregoing simple approach is not permitted any longer. Even in the case of a circular tunnel (Fig. 39) the necessary free anchor length cannot be evaluated without taking the stress-strain law and the dilation of the rock into account.

d) Effect of prestressing the anchors

When prestressed anchors are used (Fig. 48), two effects are observed (Fig. 49):

a) The tensile stress in the steel tendon is nearly constant as long as the external force Z is inferior to the prestressing force V_o. The displacements of the anchor plate and the fatigue of the tendon are negligible even at varying values of Z.

b) The rock situated between anchor plate and bond length remains always compressed; its disintegration is effectively prevented. The compressive strength parallel to the surface is considerably increased even by the comparatively small compression exerted by the anchors.

e) Evaluation of the admissible anchor load

The admissible load of an anchor is obtained by dividing its failure load by an appropriate factor of safety. Generally two proofs are necessary:

ea) Safety of the steel tendon

Indicated against reaching either the stress at failure or at the begin of plastic deformation.

eb) Safety of the bond length

Indicated against reaching the pulling out force evaluated by tests. In plastic rock types, also the long term behaviour must be considered (Fig. 50); the safety is indicated against that force Z where α reaches a given value.

At the site, a certain control is done for each anchor when pulling it at a force superior to the admissible force, before the transfer of load to the head of the anchor.

f) Evaluation of the transfer load

The transfer load, i.e. the load at which the anchor is blocked after tensioning is not necessarily equal to the admissible load. For a best possible prevention of rock disintegration, the transfer load should reach the admissible one. Anchoring in swelling rock or, e.g., the stepwise excavation of a cut in rock (Fig. 35, 1 to 4) however implicate a further displacement of the anchor head, after having blocked the anchor. In order to take account of the resulting increase of the anchor force, the transfer load should be reduced and provisions have to be taken to control and to readjust the anchor force, at certain intervals.

g) Factor of safety obtained by anchoring

Due to the difficulty to give a universally convenient definition of the "factor of safety" in rock engineering, restriction is made to the following considerations:

a) The safety of the rock structure obtained by anchoring is not that of the single anchors, treated at the occasion of defining the admissible anchor load. A variation of the applied anchor forces influences the safety of the structure in a more or less important proportion (Fig. 51) depending on the given situation.

b) The global stability of the rock structure does not depend explicitly on the parameters of anchoring. In fact, it depends on the arrangement of the anchors as the stability examination·is made for a potential slip line situated behind them.

c) The increase in safety obtained by the prevention of rock disintegration can hardly be expressed numerically, but may be very important.

3) TECHNOLOGY OF ANCHORING (SELECTED CHAPTERS)

a) Introduction

In the context of the present lecture, only brief notices concerning anchor technology are given.

b) Anchor system

In order to have an anchor with well defined design parameters, the anchor system shall allow for discerning not only the free steel length from the bond length of the tendon, but also the free anchor length from the bond length of the anchor (Fig. 52). A packer separating the two latter parts permits to grout only the bond length and to do the grouting with high pressure.

Generally, the annular space between the sheath of the anchor and the borehole wall is comparably small so that the part of the anchor force transmitted into it when there is no packer and when the annular space is also filled with grout, is not of great importance. The fact remains, however, that generally no high pressure grouting can be done when the packer is at the borehole mouth.

c) Secondary grouting

In order to protect also the free steel length of permanent anchors against corrosion, the annular space between the tendon and the sheath is filled with an anticorrosive medium.

It is grouted with a cement mix after a certain stabilization of the rock displacements, the anchor is then blocked; no control or readjustment of the anchor force is possible any longer.

Filling this space with an elastic or plastic medium allows the free steel length to remain freely movable. This advantage must, however, be paid by higher costs.

d) Protection against corrosion

As the anchors are not made from pure iron, electromechanical potentials arise at the steel surface leading to electrolytical dissolution in the presence of a dissociated medium. This phenomenon, i.e. the corrosion, is the enemy number one of permanent anchors. The intensity of the corrosion depends on the magnitude of the electro-chemical potential, on the pH-value of the surrounding medium (Fig. 53) and on the tension of the steel.

The commonly used protective measures belong therefore to three groups:
a) Isolation of the steel from the dissociated medium: no electrolytic effect can occur. Examples: bitumen, water free organic resins, nitrogen atmosphere, mineral greases.
b) Artificially negative electro-chemical potential: cathodic protection.

c) Choice of a dissociated medium with pH $\cong$ 10: cement grout, elastomeres (appropriate aqueous organic resins).

4) EXAMPLE

a) Situation (Fig. 54)

A vertical shaft of 25 m diameter and 50 m depth shall be built in the bottom of a valley. The rock is siltstone (density γ = 2,7 t/m^3) dipping 57° northwards. Vertical joints striking N-S and the bedding planes are exceptionally planar and frequently filled with a clayey silt showing no cohesion and g = 20° internal friction.

Groundwater needs not to be taken into consideration. Flat jack tests performed at 50 m depth in a test gallery showed a horizontal stress, perpendicularly to the joint of 1,7 - times of the overburden.

b) Scope

The wedge shown in Fig. 54 risks to slide into the shaft and shall therefore be stabilized by an anchor system.

c) Partial safety factors

Defined accordingly to the degree of uncertainty of the different parameters.

Density of the rock: ν_γ = 1,00
Internal friction of joint filling material: ν_ρ = 1,20
Horizontal stress: ν_{σ_h} = 2,5

d) Stability of the wedge

Active force: $G . \sin \vartheta$
Reaction forces: $(G . \cos \vartheta + 2 R_s + E . \sin \vartheta) \, tg\rho + E . \cos \vartheta$
Condition of equilibrium:

$$G . \sin \vartheta - \left(G . \cos \vartheta + 2 \frac{R_s}{2,5} + E . \sin \vartheta \right) \frac{tg\rho}{1,2} - E . \cos \vartheta = 0$$

$$E = \frac{G \cdot \sin\vartheta - \left(G \cdot \cos\vartheta + 0{,}8 \; R_s\right) \frac{tg\rho}{1{,}2}}{\cos\vartheta + \sin\vartheta \cdot \frac{tg\rho}{1{,}2}} = \frac{0{,}68 \; G - 0{,}24 \; R_s}{0{,}80} = 24700 \; t$$

with: $\sin\vartheta = 0{,}84$; $\cos\vartheta = 0{,}54$; $tg\rho = 0{,}36$

$G = 63840 \; t$

$R_s = 98060 \; t$

Assumption: uniformly distributed support $e = E / \phi \cdot D = 19{,}7 \; t/m^2$

e) **Support system**

Creation of a virtual arch with 3 hinges in the rock, loaded by the support pressure e. Anchors serve to keep the resulting force acting in the arch at a max. inclination to the joint normal of 20° (Fig. 55).

Horizontal reaction H of the arch : $H = e \cdot \phi^2 / 8 \; f = 290 \; t/m$

Evaluation of the necessary anchor forces:

$$A \cos\xi + A \cdot \sin\xi \cdot tg\rho + H \cdot tg\rho = E = e \cdot x$$

$$A = \frac{E - H \cdot tg\rho}{\cos\xi + \sin\xi \cdot tg\rho} = 20{,}5 \; x - 108 \quad \text{with } \xi = 45°$$

$$A = 0 \text{ for } x_0 = \frac{108}{20{,}5} = 5{,}3 \; m$$

$$\text{for } x = 12{,}5 \; m : A_{12.5} = 20{,}5 \cdot 12{,}5 - 108 = 148 \; t/m$$

This anchor force is provided by three lines of anchors distributed as shown in Fig. 55.

f) Choice of the anchors

Steel section 6 ϕ 16, St 125/140

For an admissible steel stress of σ_{adm} = 0,55 $\cdot$ σ_f = 0,55 $\cdot$ 14,0 = 7,7 t/cm^2

the admissible achor load is A_{adm} = 6 $\cdot$ 2,0 $\cdot$ 7,7 = 92,5 t

the vertical distance between the anchors is a_{vert} = 92,5/48,3 = 1,9 m

Total number of anchors N = 6 x 50/1,9 = 158

Total capacity of anchors: A = 158 $\cdot$ 92,5 = 14 600 t

Remember: necessary support force without arching effect of the rock:

$$E = 24\ 700\ t\ !$$

g) Length of the anchors

a) Bond length begins at the joint which is tangent to the shaft.

b) Value of bond length chosen to 5 m, i.e. shear force per unit length = 18,5 t/m or shear stress for a borehole with ϕ = 12 cm: $\tau \cong 50$ t/m^2. As this value is relatively small, a certain reduction of the bond length should be possible.

D) DRAINAGE

Refer to the lectures given by M. Cl. Louis.

REFERENCES

[1]   Müller, L.: Der Felsbau, Bd. 1; Enke Verlag, Stuttgart, 1963

[2]   Cambefort, H.: Géotechnique de l'ingénieur; Eyrolles, Paris, 1971

[3]   Müller, L.: Tunnel- und Stollenbau; Lectures, University of Karlsruhe

[4]   Lurgeon, M.: Barrage et géologie; Dunod, Paris, 1933

[5]   Talobre, J.: La mécanique des roches appliquée aux Travaux Publics; Dunod, Paris, 1957

[6]   Cambefort, H.: Injection des Sols; Eyerolles, Paris, 1964

[7]   Rescher, O.J.: Die Anwendung des Gefrierverfahrens beim Ausbau eines Stollens in einer schwierigen Gebirgsstrecke; Rock Mech., Suppl. I, Springer Wien, 1970

[8]   Ostermayer, H. and Werner, H.-U.: Neue Erkenntnisse und Entwicklungs-tendenzen in der Verkerungstechnik; Deutsche Baugrundtagung, Stuttgart, 1972

[9]   Comte, Ch.: Technologie des Tirants; Inst, for Engineering Research, Found. Kollbrunner/Rodio, no. 17, Leemann Verlag, Zürich, 1971

[10] Different authors: Anchorages especially in soft ground; Speciality Session no 15, VIIth Int. Conf. on SMFE, Mexico, 1969

[11] Birkenmaier, M.: Vorgespannte Felsanker; SBZ, 71. Jg. Nr. 47, 1953

[12] Lang, Th.: Theory and practice of rock bolting; Trans. Amer. Inst. Mining Met. Engrs., Mining Div., Vol. 223, 1962

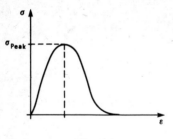

Fig. 1a

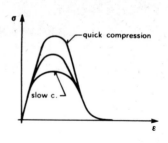

Fig. 1b

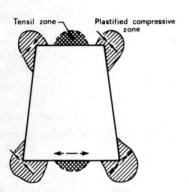

Fig. 2

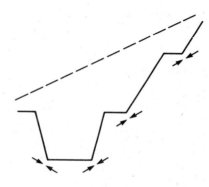

Fig. 3

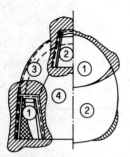

Fig. 4   Fig. 5

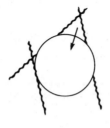

Fig. 6

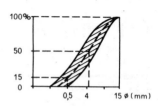

Fig. 7

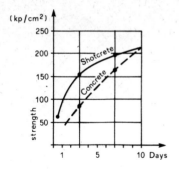

Fig. 8

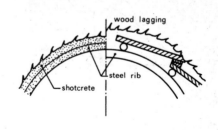

Fig. 9

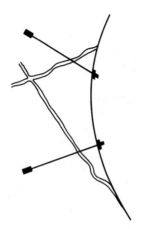

Fig. 10

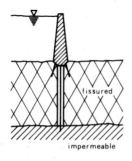

Fig. 11

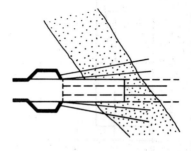

Fig. 12

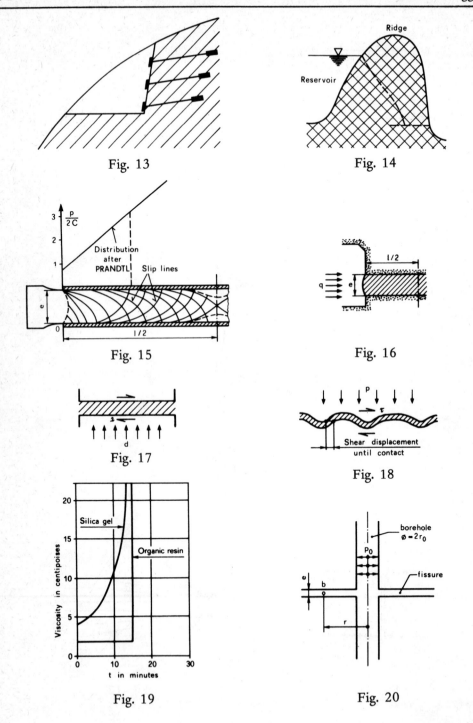

Fig. 13

Fig. 14

Fig. 15

Fig. 16

Fig. 17

Fig. 18

Fig. 19

Fig. 20

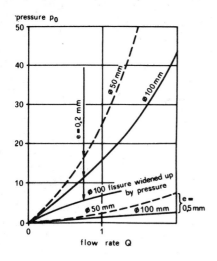

Fig. 21

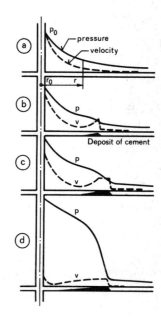

Fig. 22

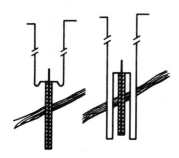

Fig. 23

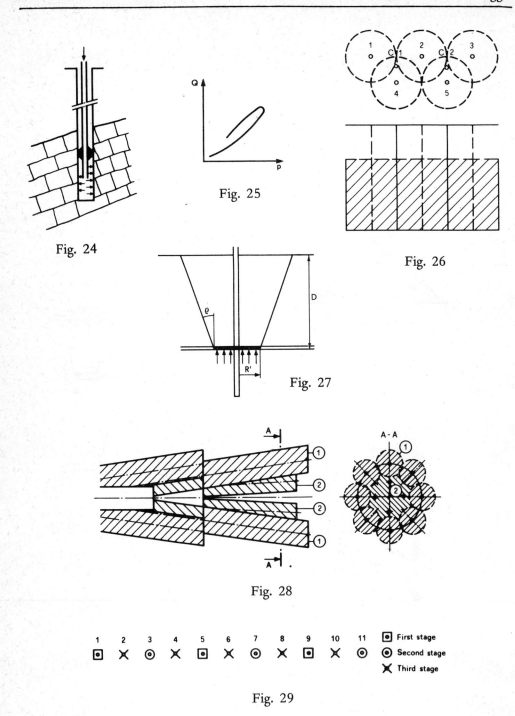

Fig. 25

Fig. 24

Fig. 26

Fig. 27

Fig. 28

1   2   3   4   5   6   7   8   9   10   11   First stage
                                                 Second stage
                                                 Third stage

Fig. 29

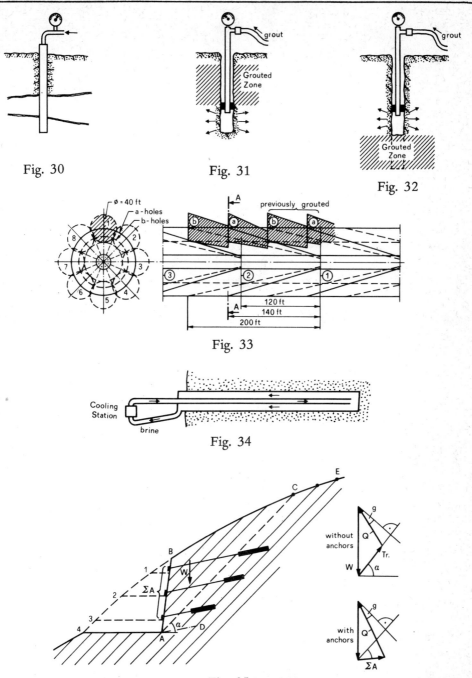

Fig. 30

Fig. 31

Fig. 32

Fig. 33

Fig. 34

Fig. 35

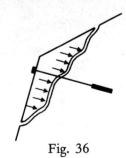

Fig. 36

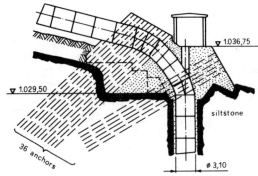

Fig. 37

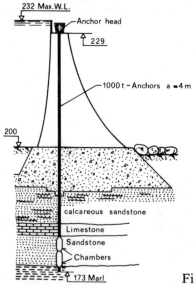

Fig. 38

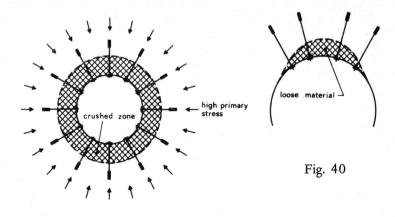

crushed zone
high primary stress

Fig. 39

loose material

Fig. 40

L1
ρ
L
L2=Lmin
ρ*
D
D=const
N
ρ
G
A1=Amin
A
A2
Assumed price of the anchor P=C.A.L.
C=Price per m and ton
A=Anchor force [t]
L=Anchor lenght [m]
P1/P2 = C A1 L1 / C A2 L2 = 1
Minimal Price
P = C·A2L2 sin ρ 1/(sin ρ*·cos (ρ*−ρ))
for ρ* = (90+p)/2 → Pmin = C·A2·L2·2 sin ρ/(1+sin ρ)

Fig. 41

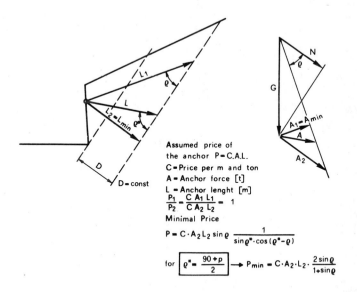

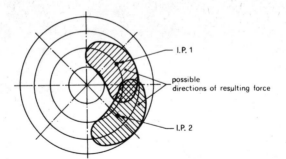

Fig. 42

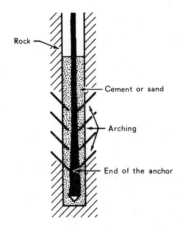

Fig. 43

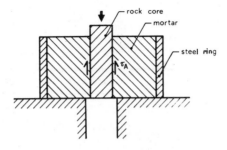

Fig. 44

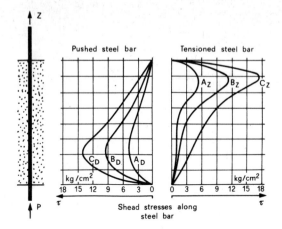

Fig. 45

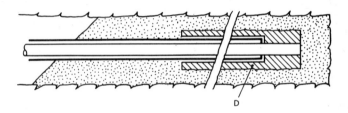

Fig. 46

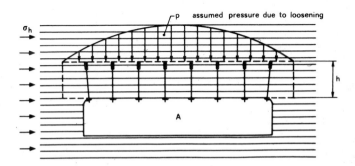

Fig. 47

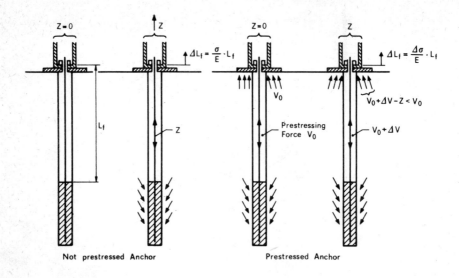

Fig. 48

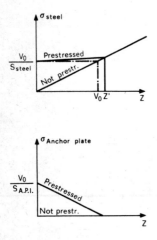

Fig. 49

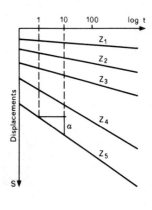

Fig. 50

a)

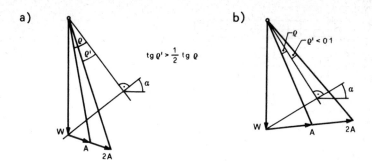

$$tg\, \varrho' > \frac{1}{2}\, tg\, \varrho$$

b)

$\varrho' < 01$

Fig. 51

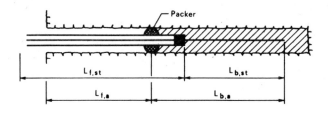

Fig. 52

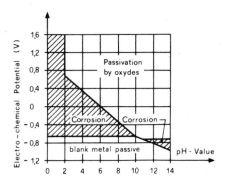

Fig. 53

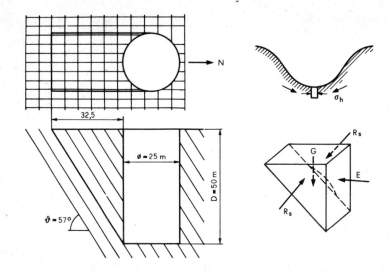

Fig. 54

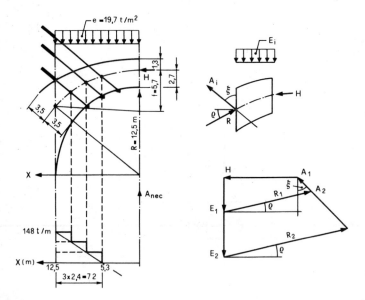

Fig. 55

UNDERGROUND OPENINGS — PRINCIPLES OF DESIGN

P. EGGER
Abteilung Felsmechanik
Universität Karlsruhe

---

* All figures quoted in the text are at the end of the lecture.

I INTRODUCTION

For a long period the design of underground openings has been almost exclusively a field of the experienced practician. The scientific support by furnishing appropriate theories was poor and restricted to — more or less — thumb rules. This is due to the fact that the parameters involved with the design of underground works are complex. Most of them are difficult and onerous to evaluate, and to a certain extent it is not yet possible (even nowadays) to take them properly into account in the stability analyses.

II DESIGN PARAMETERS

A) ROCK MASS

It is obvious that for a construction in rock, the main set of parameters concerns the mechanical conditions of the rock mass in its natural state and its behaviour during and after construction.

1) Geological conditions

The knowledge about the geological situation is indispensable for judging the general feasibility of a proposed construction site. The existence of e.g. major faults or thrusts in a parameter influencing strongly the general design of underground works.

2) Mechanical properties

No reasonable static approach to the evaluation of the stability of a tunnel is possible without knowig certain mechanical properties of the rock mass. The most fundamental properties for tunnelling are the complete stress-strain-law (including the dilation behavior) and the time-dependence of the rock properties.

As the rock mass is generally heterogeneous, anisotropic and crossed by discontinuity planes, the evaluation of the mechanical properties may become very complex, and scale effects have also to be taken into account.

It is most important to note that the properties of the rock **mass** must be considered, not those of the intact rock **element**.

3) Primary stress field

The primary stress field, acting in the rock mass before the begin of construction works may have almost the same incidence on the stability of a rock cavity as the mechanical properties of the rock mass. Generally it cannot be evaluated by theoretical considerations because it depends strongly on the local geological history (sedimentation, tectonic and seismic effects, erosion a.s.o.) which is never completely known.

4) Hydraulic conditions

In a tunnel design, also the hydraulic conditions encountered at the site have to be considered: Hydrodynamic pressure and flow rate expected during the construction, hydrostatic pressure acting on the final   lining.

B) UNDERGROUND WORKS
1) General layout

In the design of underground works e.g. for a hydroelectric power plant, a certain number of large cavities and smaller connection tunnels and shafts are needed. Their mutual arrangement, i.e. the width and shape of the remaining rock pillars is an important factor for the overall stability.

2) Orientation

Due to the anisotropy of the rock properties and of the stress field, the stability of an underground cavity is also influenced by its orientation.

3) Shape

The shape of the opening has a notable influence on the risk of encountering difficulties during construction, i.e. on its stability.

C) CONSTRUCITON METHOD AND SEQUENCE

The usual stability analyses are unable to take the incidence of the adopted construction method and sequence properly into account. The experience teaches, however, that the safety of a tunnel construction is strongly influenced by these parameters.

1) Methods of excavation

The modifications of the rock qualities near the perimeter of the tunnel due to the excavation method (blasting, tunnel mole) are susceptible to have an important influence on the stability of the excavation.

2) Sequence of excavation

The repeated redistributions of stresses and strains around a tunnel when it is excavated by partial sections may influence the stability very unfavourably.

3) Support

Last not least, the kind of the adopted tunnel support as well as the moment of its installation must be taken account of in the design.

III TENTATIVE GUIDELINE FOR THE DESIGN

From this brief review of the embarassingly high number of complex parameters influencing the stability of an underground cavity it becomes evident that even the most refined design based on the best available results of investigations and tests, is not much more than a rough approach to the problem. A design worked out before the begin of the construction should, therefore, never be considered as definitive, but as preliminary and subject to modifications according to the experiences acquired during the construction.

In the following, a tentative guideline shall be given how to approach the design of underground openings.

A) GEOLOGICAL INVESTIGATIONS

The first informations about the general suitability of a site are obtained by geological studies: Surface mapping, interpretation of cored boreholes etc. Special attention shall be paid to detect the geological features particularly relevant to the construction such as the orientation of discontinuity planes encountered as sets or as individuals (faults or similar features). These investigations should also include a hydrogeological campaign. A first classification of rock types can be given by the engineering geologist.

B) GEOTECHNICAL INVESTIGATIONS

The scope of the geotechnical investigations is to replace the qualitative description of the site conditions given by the geologist by a quantitative one. In-situ and laboratory tests are performed to obtain informations about the design parameters referring to the rock.

1) In-situ-tests

The in-situ-tests are appropriate to evaluate the primary stress field and the mechanical behaviour of the rock mass in place. The disturbances of the original conditions created necessarily by the construction of test galleries or similar render the correct interpretation of the test results somehow difficult.

In order to obtain representative results for the behaviour of the rock mass, the rock volume interested by the tests must contain a sufficiently high number of rock elements limited by discontinuity planes.

2) Laboratory tests

Laboratory tests yield the physical and mechanical properties of the rock matrix and, under certain precautions, the behaviour of joints.

C) INTERPRETATION OF EXPLORATORY WORKS AND TESTS

The variety of data and test results obtained by the investigations must be prepared and somehow schematized in order to yield a useful input for the design properly speaking.

1) Evaluation of rock classes

Based on the classification given by the engineering geologist and on the test results, a refined classification is worked out containing not more than about 6 rock classes, in the general case.

2) Evaluation of mechanical rock parameters

The average mechanical parameters are assigned to each rock class, noting the scatter.

3) Evaluation of primary stress field

Based on the results of the in-situ tests which scatter frequently within a wide range, and of static considerations, the most probable primary stress field is evaluated. It is convenient in many cases to give an upper and a lower bound for the horizontal stress.

4) Evaluation of design criteria

Before starting the design, some principal criteria must be given. They include the choice of safety factors, depending on the nature of the planned work as

well as on the scatter of the available data. Furthermore, criteria referring to the
economy and to the **organization** of the site have to be defined in time. The most
elegant solution is not necessarily the most economic one.

D) DESIGN OF THE UNDERGROUND WORKS
1) General concepts
a) Adaptation to the given situation
Contrarily to surface constructions (e.g. house building), in underground works a reasonable and economic designer will never try to simply carry a
design which proved to be convenient for one site, over to another one. Each case is
an individual one, requiring special attention.

b) Conservation of rock qualities
As it is outlined in the lectures on "Rock Stabilization", the systematic
attempt to conserve the original qualities of the rock mass begins at the stage of
design. Its importance for the safety and economy of the works can hardly be
overestimated.

2) Desgin (properly speaking)
a) First stage
Beginning with the great features the design will first define the general
arrangement of the works. Besides, the planning of the site is worked out and the
most appropriate method of construction is retained.

b) Second stage
Then, the detailed design comprises the definition of the shape of the
underground opening (in the final stage, but also in intermediate stages), considering
cross sections and longitudinal sections and containing stability studies and the
evaluation of the rock support and of the final lining. All those different phases of
work being in a close interdependence, the final solution is the result of a series of
trial and error studies which lead to successive refinements of the first draft.

c) Remarks
α) **Geology**: If the geological conditions on the site show important individual
features such as major faults, their implications to the stability of the opening
have to be assessed specially. Frequently, a rough static analysis of rigid bodies

is sufficient to show whether or not there is e.g. a risk of squeezing out a rock wedge limited by a number of joint planes.

β) **Size effect:** The usual procedures of analysis do not take account of the size effect. They yield, for example, a certain value for the circumferential normal stress acting near the surface of a tunnel situated in great depth, independently of the value of its radius. The experience shows, however, that a small tunnel may be stable in a given rock whereas a large excavation may be unstable. In order to allow for this size effect a certain reduction of the rock parameters should be carried out for large excavations.

γ) **Time dependent behavior of the rock:** It is a matter of experience that the rock needs a certain time to deform under applied stresses. Uncontrolled deformation is frequently accompanied by a disintegration. Again, this effect can yet hardly be incorporated to a satisfying degree in static analyses, but must not be underestimated by the designer and the constructing engineer.

E) STABILITY STUDIES
1) Basic assumptions

Even the most refined methods to perform stability studies are not better than the idealization carried out in order to make the complex variety of natural conditions accessible to an analysis.

a) Two- or threedimensional

Generally, mathematical analyses are performed for the case of plain strain, for reasons of mathematical commodity. This is a realistic approach for assessing the stability of a tunnel in its final stage or of the central part of an elongated cavity. It is, however, of a very restricted value for the examination of the stability near the tunnel face during excavation or near the ends of a cavity. The influence of the third dimension becomes prevailing there (combined with that of time dependent behaviour of the rock) and is the reason why tunnelling in a rock which needs support is actually possible.

Whereas the consideration of the plane is conservative for concave rock shapes, it is unsafe for the case of convex rock shapes such as intersections of two galleries.

b) Adopted Material Laws

As an underground opening is a hole in a large rock mass, the mechanical behaviour of the rock is of highest importance for the stability of the opening.

α) **Stress-strain law:** The stress-strain laws of the rock mass such as evaluated by testing are always more or less radically simplified for the usual static analyses. In mathematical calculations, linear elasticity followed or not by plastic behaviour is the most usual assumption. Methods which allow for non-linear elasticity and for a progressive disintegration after the peak stress, taking also the true dilation behaviour into account are not yet generally used.

      The adopted **failure criterion** separating the pre-failure from the post-failure-zone and giving the peak strength is usually simplified, too. Frequently, separate simple failure criteria are indicated for the matrix and for the joints.

      In model tests with equivalent material it is, however, possible to reproduce quite well some aspects of the complex behaviour of the natural rock mass.

β) **Time dependence:** Viscoelastic or viscoplastic phenomena exhibited by the rock have been attempted to reproduce in mathematical analyses at the cost of other important rock properties. It is obvious that the introduction of a new dimension risks to multiply in an explosive manner the amount of calculation work and computer time.

2) Types of stability studies
a) Mathematical studies

      The most commonly used stability studies are mathematical analyses of different degrees of refinement: equilibrium of rigid bodies, assumed loads acting on a support structure, consideration of a hole in the rock mass by analytical or numerical methods.

α) **Equilibrium of rigid bodies:** In certain cases with few kinematic restraints, it is sufficient to consider the equilibrium of forces acting on a rigid body. For example the stability of a rock wedge limited by the tunnel and by joints may be evaluated in this way.

β) **Assumptions of external loads:** Civil engineers familiar with the statics of concrete structures tend frequently to transmit the ways of consideration which are common in this field, to underground works. The idea is to assume external loads ("rock pressure") acting on a tunnel support, the latter being considered as an arch bedded elastically or not over a certain portion of its length. The resulting moments and forces (normal, shear) allow then to dimension the support.

The crucial problems are, however, the evaluation of the "external" loads, both vertical and horizontal, and that of the reaction modulus of the elastic bedding. More or less arbitrary assumptions for these parameters leed necessarily to a wide scatter of the results of calculations.

In certain cases, however, this one-dimensional analysis yields certainly valuable results, for example for assessing the sollicitations of the final lining in a tunnel situated near the surface.

γ) **Interaction of rock mass and structure**: The way out from the impasse where the one-dimensional analysis leads generally, is the extension of the analysis to two or three dimensions. As the latter complicates enormously the calculation work, most of the common methods assume plane strain.

In order to obtain valuable results, the different construction stages must be taken into account; of particular importance are the questions at what degree of rock deformations the support is built in and what reaction forces this support develops at a further displacement of the rock surface.

γ1) **Analytical procedures**: permit a systematic examination of the influence on the stress and strain fields of certain parameters such as the magnitude of horizontal stresses, the Poisson's ratio, the disintegration characteristics in the Post-Failure region or the shape of the opening. Thus, a clear assessment of the impact exerted by a variation of these parameters can be given without too much calculation work. On the other hand, a series of idealizations are necessary, the rock being modelled by a homogeneous, frequently also isotropic medium without discontinuities, the shape of the opening is described by a simple geometrical function (e.g. circle, ellipse).

= **Elastic rock**: The problem of the hole in an infinite elastic plate has been treated repeatedly since Lamé, investigating circular and elliptical holes in an isotropic or anisotropic stress field.

One most instructive result is the distribution of the circumferential stress along the perimeter of the hole: important concentrations of compressive stress or tensile stresses are possible according to the choice of parameters.

= **Elastic-ideally plastic rock**: This type of analysis allows for a failure criterion of the rock and assumes an ideally plastic Post-Failure-behaviour. The high stress concentrations known from the elastic analysis are essentially reduced (see Fenner, Talobre, Kastner) due to the appearance of a plastified zone; the high circumferential stresses are pushed from the tunnel surface inwards into the rock. The usual methods require axial symmetry and a volume constant plastic flow; they give the stress field but do not care about the strain field.

= Elastic-plastic rock taking Post-Failure-behaviour into account: A progressive disintegration of the rock after having reached the peak strength is a matter of experience. It may be rather abrupt for brittle rocks or little pronounced for rather ductile rock types. Knowing the shape of the disintegration curves and the magnitude of dilation of the rock from complete compression tests (see Rummel), it is possible to evaluate the stress and strain fields in the axisymmetric case (see Egger). Depending on the magnitude of primary stresses, on the failure criterion and on the characteristics of the Post-Failure-behaviour, different stability cases can occur: either the need of an immediate tunnel support of relatively high resistance in order to hinder too great deformations; or the establishment of a new equilibrium without support; this equilibrium being more or less unstable in time, it must be preserved by a support which prevents further rock displacements due to perturbation forces (e.g. blasting).

γ2) **Numerical procedures:** The inconvenient of the analytical procedures such as restrictions concerning the shape of the opening, anisotropy of the rock and the stress field can be overcome effectively since the widespread introduction of the computer into engineering calculus. Finite element and finite difference techniques are now available in versions appropriate to rock and soil engineering. These methods are generally based on elastic, or elastoplastic rock behaviour; in the former case, even three-dimensional computations are possible with a reasonable expenditure.

Post-Failure-characteristics are not yet allowed for in the available programs.

b) Model tests

Certain difficulties against which the mathematical methods hit because of the too large number of operations, can be avoided by model testing. The main problem is to choose the most appropriate model law as it is impossible to take account of all parameters describing the behaviour of the rock.

α) **Mechanical models with equivalent material:** When for a specified case the selection of the most important mechanical properties is done, the rock can be modelled by an "equivalent" material exhibiting these same properties multiplied by a scale factor dependent on the adopted model law (see Oberti, Fumagalli). For example, the influence of discontinuities (bedding, joints, faults) on the stability of an underground cavity can be shown clearly (see Sharma).

β) **Different materials:** In order to avoid the onerous working out of equivalent model material, simpler tests are frequently carried out which show quickly the overall behaviour of the structure, but are of rather qualitative type because of the discrepancy between the material laws of the rock and the model material. Photoelastic studies or simple feature models may be quoted here.

c) Prototype studies

Despite the incontestable merits of mathematical analyses and model studies, complete informations about the anticipated behaviour of an underground opening are only obtained by the construction and monitoring of a prototype; this may be, e.g., a portion of tunnel about three diameters long.

The studies undertaken on a prototype are the only ones to avoid the impact of the scale factor and to assess reasonably the influence of the construction method. Though the construction of a prototype is expensive, absolutely seen, it can be incorporated in the global project forming thus part of it; moreover, the multitude of obtained informations leads, by experience, to a sounder assessment of construction costs by the contractors and, in most cases, to a less expensive tender.

IV CONSTRUCTION (Selected Chapters)

A) INTRODUCTION

In the context of this lecture, only some thoughts concerning the excavation and temporary support, as well as the central role of monitoring are presented.

B) STAND—UP TIMES

By experience, Lauffer found out a relation between the length (or width) of an unsupported excavation and the time at which slabbing or falling down of rock begins ("stand-up time"), which in a double-logarithmic scale is linear. According to the rock types, the straight lines are positioned on the diagram, permitting thus an empirical rock classification.

Although empirical, this rock classification is most useful for the construction. It gives valuable informations about the admissible excavation length and the time available for build in the temporary support.

C) TEMPORARY SUPPORT AND FINAL LINING

The classical support methods using frames (or ribs) plus a wooden lagging furnish only a very localized support of the rock surface, Between two adjacent lags, the rock is free to move inwards to a certain extent, what favours its progressive deterioration. The support is, indeed, not much more than a temporary protection of the labourers. A final lining is needed in order to carry the loads exerted by loosened rock zones.

A modern support system such as used by the New Austrian Tunnelling Method furnishes, by the projection of shotcrete quickly after the excavation, a continuous support which is in intimate contact with the rock. By the use of prestressed rock bolts, this intimate contact is yet intensified and the rock is enabled to sustain considerably increased stresses in the tangential direction. On the other hand, this type of support is not rigid but follows the general trend of displacements without being overstressed. In fact, this is more than a temporary support and in many cases no separate final lining is necessary (e.g. power station of Waldeck).

D) MONITORING

In order to overcome the lack in knowledge during the design stage, the results of the adopted methods (excavation, support) have to be carefully controlled.

1) Displacement

The easiest way of controlling the behaviour of an underground cavity is to measure the relative displacements of points situated on its surface. This is completed by levelling a series of points situated at the key of the arch. Extensometers and deflectometers yield valuable informations about the strain field in the rock mass.

2) Stresses

The contact pressure between rock and support as well as the stresses acting in the concrete are given by conveniently installed pressure cells.

3) Interpretation

Whereas the results of pressure measurements are directly applicable for the evaluation of the safety factor inherent to the concrete lining, the displacement curves yield the possibility to judge the overall stability. They show clearly whether or not the rock displacements tend to stabilize after the support has been built in. If no stabilization occurs, this is a signal of alarm and measures to reinforce the support have to be undertaken.

When speaking about a "safety factor" particular attention must be paid at its respective definition, as no generally accepted definition exists. It is a matter of experience that a tunnel may be unstable without any support, but it may be stabilized by a support even when this support has yielded. Thus, the only fact that yielding failure occurs in the support does not mean necessarily that the tunnel collapses.

REFERENCES

[1]   Müller, L.: Der Felsbau, Bd. 1; Enke-Verlag, Stuttgart, 1963

[2]   Stini J.: Tunnelbaugeologie; Springer, Wien, 1951

[3]   Laufer, H.: Gebirgsklassifizierung für den Stollenbau; Geol. u. Bauw., Jg. 24, H. 1., 1958

[4]   Kastner, J.: Statik des Tunnel- und Stollenbaus, Springer, Wien, 1962

[5]   Pacher, F.: Deformationsmessungen im Versuchstollen als Mittel zur Erforschung des Gebirgsverhaltens und zur Bemessung des Ausbaus; Felsmech. u. Ing.-Geol., Suppl. I, Springer, Wien, 1964

[6]   Fenner, R.: Untersuchungen zur Erkenntnis des Gebirgsdrucks; Glückauf, Jg. 74, Essen, 1938

[7]   Talobre, J.: La mécanique des roches; Dunod, Paris, 1957

[8]   Malina, H.: Berechnung von Spannungsumlagerungen in Fels und Boden mit der Hilfe der Elementenmethode; Veröff. Inst. Boden- u. Felsmech. Univ. Karlsruhe, H. 40, 1969

[9]   Goodman, R. and Dubois: Duplication of Dilatancy in Analysis of Jointed Rock; Journal SMFE, ASCE, Vol. 98, No. SM4, 1972

[10]   Rummel, F. and C. Fairhurst: Determination of the Post-Failure-Behaviour of Brittle Rock Using a Servo-controlled Festing Machine; Rock Mech., 2, Springer, Wien, 1970

[11]   Egger, P.: Einfluss des Post-Failure-Verhaltens von Fels auf den Tunnel--ausbau (unter besonderer Berücksichtigung des Ankerbaus); Veröff. Inst. Boden- u. Felsmech., Univ. Karlsruhe, H. 57, 1973

[12] Lang, T.A.: Rock Behaviour and Rock Bolt Support in Large Excavations; Symp. Underground Power Stations, ASCE, Power Div., New York, 1957

[13] Rabcevicz, L. von: Bemessung von Holraumbauten; Felsmech.- u. Ing.- Geol., Suppl. II, Springer, Wien, 1965

[14] Rescher, O.J., K.H. Abraham, F. Bräutigam and A. Pahl: Ein Kavernenbau mit Ankerung und Spritzbeton unter Berücksichtigung der geo- mechanischen Bedingungen; Rock Mech., Suppl. 2, Springer, Wien, 1973

[15] Fumagalli, E.: Statical and Geomechanical Models; Springer, Wien, 1973

[16] Sharma, B.: Das Verhalten des Gebirges in der Umgebung von Lehnen- tunneln (Projekt C.5); Sonderforschungsbereich 77, Univ. Karlsruhe, Jahresbericht 1972, Karlsruhe 1973

[17] Different authors: Underground Rock Chambers; Symp. ASCE, Phoenix, 1971

ROCK HYDRAULICS

C. LOUIS
Bureau de Recherches Géologiques et Minières
Service Géologique National
Orléans Cédex

---

* All figures quoted in the text are at the end of the lecture.

SUMMARY

This report gives some new considerations on rock hydraulics with application in the triple field of civil, mining and petroleum engineering. The rock medium is assumed to be a jointed medium with a low conductivity of the rock matrix.

After a brief analysis of the hydraulic characteristics of rock masses and of the laws governing flow in fissures, both continuous and discontinuous, this report examines the different mathematical or physical models (electrical or hydraulic) wich make it possible to solve problems of three-dimensional water flow through jointed media with one or more sets of parallel fractures. Several methods are suggested, according to the nature of the rock jointing. The mathematical or physical models are elaborated with the help of the concept of directional hydraulic conductivity directly measured by a new *in situ* technique.

The theory of water-flow through jointed rock has undergone rapid advances in recent years. Computing techniques can now be applied to very large, two and three dimensional problems. However, the application of these methods is completely inadequate if no *in situ* hydraulic parameters are available. Several suggestions are given for the control of groundwater flow and the hydraulic instrumentation.

Several practical examples concerning dam foundation, slope or underground openings, illustrate the methods of solution. In particular, the problem of drainage in rock is examined. A fundamental difference is made between the drainage of the joint network and the drainage of the rock matrix. Practical cases show that the optimum direction of the drain depends essentially on the geometry and orientation of the joints. For a required draining effect, the cost of a drainage system will be notably reduced if the geometry (direction, depth, etc...) is judiciously chosen.

1. STATEMENT OF THE PROBLEM

The engineer is faced in many fields of science with problems connected with the flow of fluids through fissured media. In civil engineering, for example, water often flows through rock masses (foundations, surface or underground structures, natural sites). In hydrogeology, as well as in the field of mining and petroleum engineering, phenomena involving the circulation of fluids through rocks play a very important part.

Research workers have, for a long time, been concentrating on the study of fluid flow through porous media; even the case of heterogeneous or anisotropic media has been studied. On the other hand, jointed media, particularly fissured rocks, constitute a field that, as yet, is not well known. Basic studies have only been made in recent years, in several countries simultaneously.

The foremost aim of "ROCK HYDRAULICS" is to analyse the hydraulic properties and characteristics of fissured masses of rock, to study flow phenomena and their effects, and lastly, to explain the behaviour of rock masses under the action of groundwater.

It is at present generally admitted that fracturing plays a decisive role in rock hydraulics. In that field, the medium is taken to be anisotropically discontinuous; the fractures that break up the rock mass give water privileged paths. The word "fracture" is here taken in a wide sense; it includes all the apertures of the rock mass, whatever their geological origin: fractures located at stratification and schistosity boundaries, joints, faults, etc... A very simple calculation makes one realize that even a few thin fractures will give the rock mass very high permeability coefficients, in comparison with coefficients for the rock matrix. Therefore, masses of rock where their permeabilities – i.e. flow phenomena – are concerned, exhibit hydraulic ansisotropies due to the fact that fractures, through geological and mechanical circumstances of their formation, do not have an arbitrary orientation, but are grouped into one or several sets of plane and parallel fractures (except, perhaps, in the case of faults).

The very structure of rock masses points to the fracture as the basic element in rock hydraulics. A systematic study was bound to start out with an analysis of phenomena occurring at the scale of the fracture, artificially separated from the mass. The rock mass can, indeed, be taken to be a combination of elementary fractures from the point of view of hydraulics. The problem taken as a whole proves to be most complex, if not insoluble; it becomes more tractable as soon as one proceeds by steps, that is, by starting with the study of a simple fracture. The first step has therefore been to study the laws governing flow in a simple fissure in laminar and turbulent flow, taking into account all the parameters likely to play a part, among others, roughness, a most important factor, the geometrical shape of the fracture, and the presence of filling material.

These results by proceeding progressively, in other words by extending the laws governing flow to the system of fissures as a whole, led to the determination of the **distribution of the hydraulic potential** $\phi = Z + P/\gamma_w$ of underground water. The hydraulic potential constitutes the essence of rock hydraulics; indeed, a knowledge of its distribution makes it possible to proceed, even for three-dimensional problems, to calculations on the flow rates and to the mechanical effects of this flow.

It has often been noted that ground-water, both flowing and at rest, has a detrimental effect on the behaviour of rock masses. Underground water affects, in the first place, the stability of the mass; sometimes, experience has even shown that water can unfortunately have unsuspected consequences and can lead to catastrophe. Thus, when a rock mass is likely to be influenced by groundwater, any stability analysis **calls for a preliminary study** of the flow network within the mass.

We shall not forget that determination of the hydraulic potential in jointed media remains our main concern. In all previous studies, an effort was made to tackle the problem in three dimensions; it is only the matter of determining the distribution of the hydraulic potential that proves difficult. The two-dimensional problem can indeed be considered as solved. Apart from the graphical and numerical methods previously suggested (Louis, 1967), several types of experimental studies were evolved (Louis, Wittke 1969-70). It has of course proved necessary, in view of the simplifying hypotheses adopted during computation, to verify theoretical results by measurements on laboratory-scale models (Wittke and Louis, 1968, Wittke, 1968) as well as on piezometric measurements performed on site (Louis, 1972). A comparison between theoretical studies and measurements on the models of *in situ* generally led to favourable conclusions.

In actual practice, a large number of problems in flow through rock masses are three-dimensional. Results obtained by the solution of two-dimensional problems are often devoid of significance. Therefore, the question of the distribution of hydraulic potential must be approached three-dimensionally. A first attempt was given in study (Louis, 1967); the method unfortunately has only a restricted field of application since it only applies to a very special type of rock fracturing.

This report recalls a few basic results and deals more particularly with the determination, **in three dimensions**, of the distribution of hydraulic potential in jointed media. In order to get acceptable results, it is of course necessary to introduce **representative** information into mathematical and physical models; special emphasis has therefore been laid on determining hydraulic parameters *in situ*.

An account of different methods is given with a few practical examples.

A number of aspects are only briefly touched upon, while others, even more numerous, are not mentioned at all. The entire subject is taken up again in detail, on a perfectly general plane, in one overall study on rock hydraulics (Louis, 1974), to be published shortly. Finally, it should also be noted that Sharp's thesis (1970) contributes a number of new and most interesting elements concerning methodology and numerical techniques in the hydraulics of jointed media; so does the work done by Maini (1971) dealing more especially with *in situ* measurements.

2. LAWS GOVERNING WATER FLOW IN ROCKS

2.1 Fissureless masses

Let us, from the start, eliminate the case of a rock mass having no open fissures. In such a medium, designated as a "rock matrix", laws of flow in porous media, such as Darcy's Law for instance apply. The rock matrix generally exhibits a very low permeability, around 10^{-7} to 10^{-14} cm/s, depending on the nature of the rock. This case is of little interest, however, for on the scale here under consideration (tens or even hundreds of meters), rock masses are always jointed or fissured.

2.2 The Single Fracture

Let us, first consider the open and unfilled fracture, possibly with bridges of rock. In rock, fractures constitute channels characterized by a high value of relative roughness, k/D_h, (where k is the absolute roughness, and is represented by the height of its asperities, and D_h, the hydraulic diameter, by twice the opening of the fracture). The relative variations in the opening of the fracture are therefore most important, since they cause, during flow, a very high pressure drop coefficient (much higher than that which may be computed from Poiseuille's Law, for instance).

Laws of flow in a single fracture can be expressed as:

$$\text{Laminar or Steady Flow} : v = k_f J_f \qquad (1)$$
$$\text{Turbulent Flow} \qquad : v = k_f' J_f^\alpha \qquad (2)$$

In these expressions, v stands for the mean velocity, k_f for the **hydraulic conductivity** of the fracture, k_f' its **turbulent conductivity**, J_f for the perpendicular projection of the hydraulic gradient ($\vec{F^*} = -\overrightarrow{\text{grad}}\ \phi$) on the plane of the fracture and, finally, α for the degree of non-linearity ($\alpha = 0.5$ for completely rough turbulent flow).

For fracture flow, the transition from laminar to turbulent takes place at very low values of the Reynolds number (*) (down to 100 or even 10), decreasing as the relative roughness of the fracture increases.

Within the fracture itself, the transition form laminar flow ($\alpha = 1$) to completely rough turbulent flow ($\alpha = 0.5$) is quite progressive; the exponent slowly changes form 1 to 0.5 when the Reynolds number changes, for instance from 100 to 2000.

The hydraulic conductivities defined in relations (1), and (2) are given by the following expressions (fig. 1):

Laminar Flow:

(3)
$$k_f = \frac{\kappa g e^2}{12\ \nu\ C}$$

Turbulent Flow:
(completely rough)

(4)
$$k'_f = 4\ \kappa \sqrt{g}\ e\ \log \frac{d}{k/D_h}$$

In these expressions, g is the acceleration of gravity, κ is the degree of continuity of the fracture (ratio of the open surface and the total surface of the fracture), e is the mean width of the fracture, ν is the kinematic viscosity of the fluid, and, lastly, C and d are two coefficients which depend on the relative roughness k/D_h of the fissure (according to Louis, 1967, $C = 1 + 8.8\ (k/Dh)^{1,5}$, and d = 1.9 for relative roughness greater than 0.033; this, in general, is the case for fractures in rock).

In the case of fractures with filling, the hydraulic conductivity is equal to the permeability of the filling, on condition, of course, that this permeability be definitely higher than that of the rock matrix.

_____

(*) The Reynolds' number, defined by the relation $Re = vD_h/\nu$, in fact has a value that is extremely difficult to determine in the case of fissured rocks, since for a given type of flow it can vary enormously from one point to another along the same fracture.

2.3 Hydraulic Conductivity of a Set of Fractures

As pointed out in Paragraph 1, it is assumed that fracturing in a rock mass is made up of several sets of parallel plane fractures. To characterize the hydraulic properties of such a medium, it will be enough to know hydraulic conductivity K (laminar or turbulent) of each set of fractures. This hydraulic conductivity may be defined, as above, through the relation between the flow velocity V (flow rate in the direction of the fractures, divided by the total cross-section of the mass) and the active hydraulic gradient, as given by the relations:

$$\text{Laminar State} : \quad V = KJ \tag{5}$$
$$\text{Turbulent Flow:} \quad V = K'J^{\alpha} \tag{6}$$

The scale of the phenomenon studied is of great importance. In a given volume, individual fractures may, within their own plane, be **continuous or discontinuous**; these two cases must be studied separately:

a) Set of continuous fractures

Directional hydraulic conductivity of a set of continuous fractures follows directly from the hydraulic conductivity of the individual fractures. It is given (in laminar flow or turbulence) by the expression

$$K = \frac{e}{b} k_f + k_m \tag{7}$$

This relationship can be obtained by dividing the flow rate by the total cross-section of the rock mass. In figure 1, e stands for the mean width of the fractures, b the mean distance between them, k_f their hydraulic conductivity and k_m the permeability of the rock matrix.

In practice, k_m is very often negligible compared with the term $e/b \, k_f$. On the other hand, if there are no cracks or if they are bounded (e = 0 and k_f = 0), only the k_m term remains in relationship (7), this case corresponding to that considered in 2.1.

b) Set of discontinuous fractures

A numerical study shows very clearly that a set of continuous fractures, even when they are extremely narrow, has a very large hydraulic conductivity (a single fissure per metre with a width of 0.1 mm corresponds to a conductivity of about 10^{-4}cm/sec; with a 1 mm width and with the same frequency, the corresponding value is 0.1 cm/sec). These theoretical values are therefore noticeable greater than the ones met with in practice, although most of the time, fissures with a width greater than 1 mm do exist. The low values of the hydraulic conductivities observed in nature can be explained simply by the fact that the fractures, even when of notable width, are of limited extent. Within their own plane, the fractures are therefore discontinuous (figure 2). Within such a medium, flow is evidently anisotropic. The fractures which do not communicate "short-circuit" any flow along their direction. The fractures are at a constant potential and the circulation of water occurs through the rock matrix.

This problem, considered three-dimensionally, has been programmed on a computer to obtain the hydraulic conductivities of such media, whatever their geometric configuration. As a first approximation, it may be assumed that the hydraulic conductivity in the direction of the fissures is given by

$$(8) \qquad K = k \left\{ 1 + \frac{1}{2} \left(\frac{\ell}{L - \ell} - \frac{\ell}{L} \right) \right\} (*)$$

Whereas the transversal conductivity is equal to k_m the permeability of the rock matrix. The degree of anisotropy is therefore:

$$1 + \frac{1}{2} \left(\frac{\ell}{L - \ell} - \frac{\ell}{L} \right) .$$

It must be noted that in rock masses with discontinuous fissures, the **degree of discontinuity** $\ell/L (\ell/L = \sqrt{\kappa}$, where κ is the degree of continuity of the surface of the fissure) and the **frequency of the fissures** b/L are **the only important hydraulic parameters.** The permeability of the matrix, k_m, occurs merely as a proportionality coefficient in the hydraulic conductivities of the different set of fractures; its influence only becomes noticeable in the computation of flow rates. **The opening and the roughness of the fissures, as well as the geometry of the fracture wall have no bearing on the problem.** In these media, the flows occur partly through the rock matrix and therefore generally remain laminar.

(*) In fact, the hydraulic conductivity K also depends upon frequency of the fissures D, and the function $K/k_m = f (1/L, b/L)$ can only be determined numerically. Relation (*) is acceptable for $3 < L/b < 6$.

Work has recently been performed, independently by the author and also at the Institut Français du Pétrole, by Dupuy and Lefebvre du Prey, (1968) on analogous questions.

Remarks in this paragraph do not apply to bedding fissures which in general are continuous; furthermore in this case the hydraulic conductivity is reduced by the presence of filling (deposited by sedimentation, weathering or alteration).

3. DETERMINATION OF *in situ* HYDRAULICS PARAMETERS

The difficulties generally encountered in the study of practical groundwater flow problems in fissured media have to be underlined. The rock mass has always a very complex geometry in which the discontinuities are spread in an anisotropic and heterogeneous manner.

The use of mathematical or physical models in the study of flow phenomena within fissured rock is only justifiable if one has enough information on the parameters *in situ*. New methods are proposed to measure hyraulic parameters which are characteristic of the rock mass. First a detailed statistical analysis of the fracture system is carried out to establish its geometrical characteristics. The results of this first step are then used for *in situ* measurement of directional permeability in boreholes.

3.1 Systematic structural analysis

Survey methods of structural geology now make it possible to count and pinpoint the different sets of fissures in a rock mass and also to determine, by statistical means, their **orientation**, their **frequency** and their shape. These parameters define the geometry of the medium. This study must be undertaken from a very specific viewpoint, so that the potential water circulation in the mass may be inferred as objectively as possible. It is divided into several distinct phases:

a) A systematic survey from trenches, reconnaissance tunnels, or oriented borehole cores, where each observed structural element is described as thoroughly as possible. The use of the enclosed check list (fig. 3) has proved to be effective for quick compilation of data and ease of computer storage.

Each structural element analysed is described by a line in the check list, corresponding to a punched card for the computer. A fracture is characterized by 17 parameters : (1) = numbering, (2) = location of the survey, (3) = thickness of cover, (4) = geographical position of the surveyed point, (5) = orientation of the face at which the fracture is deserved, (6) = rock type, (7) = structural element, (8) = orientation (dip direction and dip) of the structural element, (9) = continuity, (10) = thickness, (11) = nature of the filling, (12) = extent of free opening, (13) = water flow, (14) = relaxation effects, (15) = spacing between fractures, (16) = continuity of jointing, (17) = friction angle. One column is left for miscellaneous observations.

A large number of structural elements must be surveyed in order to permit a valid statistical treatment — several hundred is a minimum.

b) A statistical treatment: The data is fed into a computer for conventional statistical treatment with representation of the fracture concentration on a Schmidt diagram. In this treatment, the fractures are given the same weight; only the number of fractures is considered. However, from the point of view of hydraulics it is certain that all fractures do not act similarly. It therefore was necessary to modify the conventional statistical analysis (based on the bulk density population), in order to obtain a better hydraulic definition of a fissured rock mass by giving some "weight" or importance to fractures appearing as the most likely to channel water. The criteria of importance from a hydraulics point of view consist of the length, continuity and opening of the fractures. Quantitatively a weight P (N) is given to each structural element (N); this weight is equal to unity and remains a constant in the conventional statistical treatment.   In the new proposed method, it varies and is an intrinsic parameter of each fracture.

Two types of weighting are proposed:

— The first emphasizes the thick and long fractures and is expressed as:

P (N) = 1 + Thickness (N) × Continuity (N); the thickness of the parameter is defined on the check list by fixing boundaries between classes. For example, this parameter may vary between 1 and 5 when the thickness of the fractures varies between 0, 1, 5, 10 and 30 millimeters. The term "continuity (N)" is determined directly in situ: It is the ratio between the length of the fracture which intersects the tunnel and the total length of the intersection if this fracture were perfectly continuous. (This term then varies between 0 and 1). At the surface it is the ratio between the length of the fracture and a reference length (this reference length can have any value).

— The second selects the hydraulically efficient fractures and is expressed as: $P(N) = 1 + \alpha \text{ (water)} + \beta \text{ (free opening)}.$

The coefficients (water) and (free opening) have been fixed **a priori** through the class boundaries as shown above. Coefficients α and β are used to give to the two scales of (water) and (free opening) an equal importance.

For practical problems the conditions $P(N) = 0$ where the free opening is smaller than a given opening, is added to clarify the diagrams by eliminating fracturing of secondary importance.

Generally the proposed weighting methods clarify the diagrams enormously by erasing the secondary features of little importance from a hydraulics viewpoint.

3.2 *In situ* hydraulic testing

3.2.1 Introduction

It is, then, impossible to try to determine the spatial distribution of all these discontinuities and the hydraulic characteristics of each one. The hydraulic parameters (for example: cross-section of fissures, roughness, filling, degree of separation or of discontinuity of the fractures, etc.), are apparently more difficult to determine *in situ*, if only because of their number (see Chapter 2). Luckily, a technique of *in situ* measurements has been perfected, which makes it possible to determine directly the **total effect** of all these different parameters through the **hydraulic conductivities K** of the different sets of fractures. It is therefore not necessary to know the detailed geometry of the fissures. The directional hydraulic conductivity K of each set of fractures is measured separately, as shown in figure 4, with the help of a single or triple hydraulic probe. In the case of a rock mass with three sets of fractures, the direction of drilling in order to test one of the sets will be chosen to be parallel to the direction of the other two joint sets. In the general case, pumping tests must be performed in three different directions. The length of the trial zones of boring should, in theory, correspond to the length of the corresponding meshes in the mathematical or physical model used to study the medium.

As described before, it is possible to determine the directions and average spacing for the different families of fractures, but it is pointed out that **no definite conclusions** (for instance, for calculation of hydraulic conductivities) can be made regarding the opening of the fractures. These are always more or less influenced, at the site, by relaxation effects, blasting, etc.. It is thus imperative to use hydraulic techniques to measure *in situ* the specific conductivity of each set of fractures.

3.2.2 Water tests-theory

Let us assume a medium in which 3 families of fractures F1, F2, F3 with total hydraulic conductivities K1, K2, K3 intersect (fig. 4).

It is possible, by observation, to determine an optimum direction for the borehole, so that this borehole intersects only one family of fractures. To test the system F1, a borehole must be drilled parallel to the intersection of F2, F3. If the medium possessed only one family of fractures, the result would be exactly representative of the conductivity for this family. But when several families intersect, there is during the test a reciprocal influence of one family on the others. That is why it is preferable to carry out tests in several boreholes perpendicular to each family.

A laminar flow in a fracture is a flow under a velocity gradient, where

$$(9) \qquad \vec{v}_f = - \overrightarrow{grad} \left\{ k_f \cdot \left(Z + \frac{P}{\gamma_w} \right) \right\}$$

with v_f = the flow velocity in the fracture

K_f = hydraulic conductivity of the fracture

Z = elevation of the point considered in the fracture

P = water pressure

$\phi = Z + P/\gamma_w$ is the hydraulic potential at each point.

Because the total hydraulic conductivity of the fracture set is proportional to the conductivity of a single fracture the above equation is also valid for the set of fractures:

$$(10) \qquad \vec{V} = - \overrightarrow{grad} \left\{ K \left(Z + \frac{P}{\gamma_w} \right) \right\}$$

with V the mean flow velocity

K the total hydraulic conductivity of the fracture set under study.

A pumping in or pumping out test from a borehole creates a flow which can be expressed by the superposition of radial flow and of a uniform flow under gravity (in the plane xoy).

$$\begin{cases} \phi - \phi_o = \dfrac{q/b}{2\pi K}\ \text{Log}\ \dfrac{r}{r_o} + (\sin\alpha) \times \\[2ex] \Psi = \dfrac{q/b}{2\pi K}\ \theta - (\sin\alpha)y \end{cases}$$

en M (r,θ) (11)

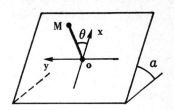

q is the flow rate in a single fracture

q/b is in fact the linear flow rate (q < o in pumping in q > o in pumping out)

r_o is the radius of the borehole

ϕ = const is a potential line

Ψ = const a stream line

During tests on the section of length L when a flow rate Q is obtained q/b will be replaced by Q/L (Q = total flow rate in the test section).

In the case of a natural flow, the gravity terms in equations (11) must be replaced by other terms taking into account the gradient Jo of the natural flow.

3.2.3 New techniques of hydraulic testing

In situ water tests are made on a large scale. They are therefore more representative than tests carried out on samples in the laboratory, but only in so far as their interpretation is possible. However, often the test cavity is not well defined and the nature of the flows is not well known; for instance there can be spherical, radial-planar or mixed flows. It is thus impossible with such tests to pretend that a correct interpretation of the results is made, because the relative importance of each flow type is unknown. In an anisotropic medium the problem is even more difficult. Therefore a new testing method had to be devised, in which the elementary permeabilities do not intervene simultaneously.

The technique used is different from the conventional Lugeon test (fig. 5) which only gives a total value of the permeability, without any allowance being made for anisotropy.

For this kind of testing the flows are indeed either spherical (in the external parts) or cylindrical (in the central part), or they are mixed. Hence, it is impossible to give from such tests a correct interpretation of the results since the relative importance of each kind of flow or, in other words, the contribution of each directional permeability value is unknown. For anisotropic media the directional permeability values are unequal. A precise interpretation of the traditional tests is not possible even for isotropic media, since the relative importance of the cylindrical and spherical flows is not known, and the equations are different for each type of flow.

Moreover, The Lugeon test does not require the borehole to have a specific direction, because it does not take into account the existing fracturing.

It is however essential to take into account the orientation of the fracturing in the interpretation of the test as shown on fig. 6.

Finally the conventional Lugeon test is carried out without piezometric control.

In the proposed new method, the test is carried out using a **triple hydraulic probe**. This eliminates all the defects inherent to the Lugeon test, that is leakage around the packers, effects at the limits of the injection section, and uncertainty in the injection pressure. The control of piezometry around the borehole is made using a point piezopermeameter. The new triple hydraulic probe to determine the directional hydraulic conductivities of jointed (or porous) media has been patented by C. Louis in France (1972). In the following more details will be given on this new technique.

In order to reduce the inconveniences mentioned in the previous section, it was necessary to develop a testing technique in which, on the one hand the type of flow was unique and completely known and where on the other hand, the elementary permeabilities did not interfere simultaneously. These conditions are all satisfied in the test-scheme of figure 7. The flow in the central section, the dimensions of which are completely known, is perfectly cylindrical with certain conditions, and the influence of the permeability parallel to the axis of the borehole has been eliminated. Such experimental conditions can be achieved by means of a hydraulic triple probe which has three testing sections limited by three or four packers. The probe is mobile i.e. it can operate at every point of a borehole, even if it has large dimensions.

Three packers only can be used (fig. 7) since the last section will be limited on the lower side by the bottom of the borehole. When the flow rate in this section is too great (for example at the upper part of a borehole) a fourth packer could be attached to the probe (fig. 8).

The tests will first be carried out at equal pressures in the three sections. Only the flow in the central section (a typical cylindrical flow) will occur in the calculations. In addition, one could carry out tests at different pressures (with flow possible between the test sections through the medium which is tested), and this would yield as a result the permeability parallel to the axis of the probe. One can obtain a wide variety of applications for the probe by thus manipulating the pressures in the various sections.

The purpose of this chapter is to give a short description of the operational method and the methods of interpretation of the results. To put it simply, piezometric measurements are required in the neighbourhood of the testing zone (at least for precise tests), and boreholes in three directions are recommended for the general case of a three-dimensional problem for a thorough study. However, in special cases, a single testing direction can suffice to determine all the hydraulic characteristics, even for three dimensional problems.

In sedimentary rock or in soil the test directions should coincide with the assumed directions of the principal permeabilities, which are generally parallel and perpendicular to the sediment beds. In jointed rock with, for instance, three systems of parallel joints, the direction of a borehole will be taken parallel to the directions of the two other joints in order to test one of the joint systems.

When there is only one borehole direction available (e.g. vertical), it is always possible to work first with equal pressures and then with different pressures in the three testing sections. In this case, a special technique and interpretation of test results must be used. But it is especially interesting to note that it is possible to determine exactly with one single test (with equal pressures), the horizontal permeability which is the only one appearing in the theory of Dupuit. It is actually possible to show that the various formulae of Dupuit which are quite often used nowadays, are correct for the calculation of the flow rates, provided that only the horizontal permeability is used in the calculations. This remains valid for stratified or multi-layered media, with the beds more or less inclined. So this observation contributes considerably to increasing the interest of the method proposed here.

3.2.4 The triple hydraulic probe

The description will be limited to the actual hydraulic probe. The accessory equipment such as surface installations, pumps, flow-meters, meters for registration of the results and piezometers acting in the testing zone will not be discussed. The hydraulic probe is schematically illustrated in a longitudinal section in fig. 9. Two models are proposed, one with hydraulic, the other with electric measurements. No dimensions are drawn in the figure because the dimensions of the probe can vary according to each particular problem (the diameter of the borehole can vary from some centimeters to several decimeters). A miniaturized version with three packers is considered in the following description (see fig. 9).

a) Probe with hydraulic measurements
— **Double tubing**, either coaxial or not (a minimum of two pipes). One of the pipes will serve for the outer sections, the other for the central section. Perforations in the pipes provide the connection between the pipes and the testing zones.

— **Three packers** (four in the case of fig. 8) preferably of a pneumatic kind which requires the installation pipes for the compressed air. Other types of packers can also be used.

The discharge ($Q_1 + Q_3$ in one pipe, Q_2 in the other according to the notations of fig. 7) will be measured on the surface and likewise the pressure. However, to account for the head loss in the pipes, the water pressures in the testing sections could be measured by means of attached pipes or pressure gauges (transducers).

b) Probe with electric measurements
— **Simple tubing:** the probe can thus function with the surface apparatus and the connection pipes used in current methods (with one single pipe).

— **Three packers** (as before).

— **Two flowmeters** with electrical transmission. The total discharge is measured on the surface, the flow $Q_2 + Q_3$ by the first flow-meter and finally Q_3 by the second flow-meter.

— **A piezometric cell** in the central section.

— **Electrical connections** between the probe and the surface.

Using the triple hydraulic probe with a double water tubing (fig. 9a) it is possible to carry out tests with different water pressures in the external and central sections. On the other hand, the device with the single tubing (fig. 9b) allows only tests with the same pressure in the three sections.

One prototype of the triple hydraulic probe has been manufactured in France and tested on several sites (Louis, 1972, 1973). The first case concerns the study of the flow conditions to a large open excavation (diameter 80 m, depth 25

m) in a fissured chalk aquifer in Lille, France (fig. 10a). The second practical application of this technique was carried out in connection with hydrogeological investigations at the Grand Maison dam-site in the French Alps (fig. 10b). After a first simulation of the flow in the dam abutments, a second important phase of the work was the verification in the field of the effects of the fractured system on the flow conditions. A series of boreholes were carefully placed around the dam-site and pumping tests were performed to verify the results of the theoretical model studies.

The use of such a set-up normally requires some measurements of hydraulic heads during the test in the vicinity of the testing area. For this control the use of a new device "the piezopermeameter" is recommended (fig. 11 or 12). Testing with the triple hydraulic probe with four piezometric measurements makes it possible to determine the three dimensional distribution of the anisotropic permeabilities or hydraulic conductivities.

3.2.5 The continuous borehole piezopermeameter

This instrument has been designed in conjunction with the triple probe. It can be used alone for a point (finite) piezometric measurement, or in conjunction with a test using the triple probe.

For piezometric monitoring two constraints exist (fig. 13 and 14):

— The measuring equipment must not disturb the natural flow net under study. However a borehole drilled through an aquifer short-circuits the different layers encountered.

— The information collected should be for a specific point and not integrated due to in-and out-flows of water in the measured zones.

Point values only make it possible to obtain hydraulic gradients.

It is therefore necessary to use a general packer which isolates the screened measurement cell. To facilitate measurements it is better, particularly for media of low permeability, that the volume of water necessary to obtain the measurement be very small. Furthermore, the response time of the system must be rapid. In this method, the measurement cell is saturated by injection of water at a pressure close to that to be measured (air removed by bleed valve). The final pressure, reached after a very long delay can be estimated by two relaxations close to the pressure assumed to exist in the mass. It is the average of the upper and lower values when the dampenings are identical during an increase or a decrease in pressure (fig. 15).

It is worth noting furthermore that the rate of variation of the value to be measured and the shape of the damping curve permit an evaluation of the order of magnitude of permeability within the measured zone.

This technique has been tested with success in France by the B.R.G.M.; some details of the equipment are given on fig. 16.

The piezopermeameter alone is also useful for solving many hydraulic problems, for instance the analysis of water effect on the stability of slopes. A possibility of hydraulic monitoring of a slope is described in fig. 17. First of all, the free surface of the ground water has to be determined by means of a few short boreholes (fig. 17a). Then a piezometric log in a single borehole (perpendicular to the free surface) determines whether the medium is isotropic (by constant piezometric head in the borehole, fig. 17b), or anisotropic with good natural drainage of the slope by decreasing of the piezometric head (fig. 17c), or with bad drainage by increasing of the piezometric head (fig. 17d).

These three cases have similar boundary conditions and free surface by a steady flow. The stability of the slope is, of course, much better in case c than in case d.

The determination of the exact piezometric log gives information on the anisotropy of the medium, the flow-net and finally the stability of the slope.

3.3 Influence of state of stress

The state of stress has a very important bearing on the hydraulic characteristics of fissured media. Thus a variation of permeability may be observed as a function of depth (self weight of the mass) and under any external influence (e.g. modification of the geometry of the medium).

The influence of stress can be introduced in the expression of hydraulic conductivities or of the permeability tensor $\bar{\bar{K}}$ (x,y,z)   K $\bar{\bar{T}}$ (x,y,z) through K, the absolute modulus of permeability. During simulation of flow in a mathematical model, it then becomes possible to modify at will the hydraulic characteristics as a function of the state of stress.

The laws of variation of the parameters which enter in the expression of the hydraulic conductivity of a family of continuous fractures are unknown. They depend on the mechanical behaviour of each type of fracture. Only an experimental approach seems realistic.

— *In situ* tests (fig. 18a)

Permeability tests in a borehole at varying depths in homogeneous fissured formations show that the empirical law which described the phenomenon is most often of the type:

$$K = K_o \cdot e^{-\alpha\sigma} \text{ with } \sigma \simeq \gamma.t \qquad (12)$$

where ko = superficial permeability, or reference permeability and $\gamma.t$ = weight of overlaying formations.

— Laboratory tests

Laboratory tests can be performed on samples crossed by one fissure only (study of permeability under stress), or on cores of smaller size, in order to look at fines jointing (fig. 18). The samples can be studied under any axisymetrical stress field by use of a parameter with longitudinal flow. A study of numerous samples with different orientations is then necessary.

Exponential laws of variation of K are frequently encountered (Louis 1974). These tests give a distribution of permeabilities which are close to reality and can then be used in the simulation by mathematical models.

3.4 General discussion on water testing
3.4.1 Criticisms of the water test analysis

The main objections to the water test analysis based on potential theory are as follows:

1) Effect of the radial flow (variation of the velocity in the flow direction).
2) Turbulence effect.
3) Deformation of the medium under joint water pressure during the test (use of unrealistically high pressures).
4) Influence of K_2 or K_3 on the test in K_1 (see fig. 4).
5) Entrance loss.
6) Influence of the time and of possible unsaturated zones.

The influence of the variation of the flow velocity in the flow direction is at a maximum in completely radial flow (e.g. for $\alpha = 0$, fig. 4). The theoretical correct equation for the potential function is given by:

$$\phi = \frac{Q/L}{2\pi K} \log r - \frac{6}{5} \frac{\bar{v}^2}{2g} + \text{constant} \qquad (13)$$

where $\phi = Z + p/\gamma_w$ = piezometric head

Q/L = flow rate per length unit

K = hydraulic conductivity of the jointed rock mass

r = radius to the borehole axis

v = mean flow velocity in the fracture perpendicular to the borehole.

The second term, due to the variation of kinetic energy, can be introduced in the elemental equation (11) but in practice this term is often negligible (Wittke and Louis, 1969).

In a water test, turbulence can begin at a very small flow rate because the gradients near the borehole are extremely high. To know if the flow is laminar or turbulent it is, in practice, necessary to plot "flow rate — hydraulic gradient" for many points by increasing flow rate from zero (see next section). The medium influence of the turbulence is near the borehole and can be taken into account as follows:

(14)
$$\phi = \epsilon \frac{1}{2}\left(\frac{Q/L}{2\pi K'}\right)^2 \frac{1}{r} - \frac{v^2}{2g} + \text{constant}$$

where $\epsilon = -1$ in pumping out

$\epsilon = +1$ in injecting.

For application of equations (13) and (14) in the water test the variation of kinetic energy $V^2/2g$ between two points 1 and 2 can be given by (for radial flow):

(15)
$$\Delta\left\{\frac{V^2}{2g}\right\}_1^2 = \frac{q^2}{8g\pi}\left\{\left(\frac{1}{er}\right)^2\right\}_1^2$$

where q = flow rate per fracture $\simeq Q/N$ (N number of the fractures in the testing sectionL),

e = mean opening of the fractures.

In practice the Lugeon test is commonly used. The water pressure in this method is generally extremely high; this can cause deformation of the medium. In the normal case (fig. 13a) the test length L is intersected by many elemental joints (L $\gg$ b). In this case, applying elastic theory, the conductivity of the undeformed medium is given by:

$$(K)o = \frac{1}{\left(1 + \dfrac{\alpha p}{E_m \cdot n}\right)} \; (K)p \qquad (16)$$

where p = integrated mean water pressure between the two considered points by the determination of (K)p,

α = coefficient whose magnitude is dependant upon the lateral stresses. α = 1 if lateral stresses are neglected and 0.5 to 0.9 if they are included.

E_m = deformation moduli of the rock matrix between two successive joints.

n = e/b = joint porosity of the considered joint system.

(K)p = hydraulic conductivity of the joint system during the test (fig, 19a)

In the extreme case of one joint intersecting the test length (fig. 19b) the equation (16) does not give the correct answer for permeability because the loading zones under water pressure for figure 19b are different to those of figure 19a. This case was considered by Sabarly, 1968.

Finally, the entrance loss in a water test can be expressed as:

$$\Delta \phi = \xi \, \frac{V^2}{2g} \qquad (17)$$

where V is the flow velocity in the joint directly near the borehole.

The coefficient ξ is in the case of a water test approximately 0.5. For 10 m/s flow velocity this loss is roughly 2.5 m head of water. This loss may be allowed for special cases.

3.4.2 Practical procedure
The water test is commonly used in practice. To obtain meaningful results from this test it is necessary, particularly for jointed rock, to observe some fundamental rules.

a) The borehole direction
In a rock mass with three joint systems the optimum hole direction for testing one joint system is parallel to the other two systems.

If the extent of jointing is very large and irregular, the medium must be considered as a continuum and a conventional soil mechanics test may be carried out, bearing in mind that the scale of the test must be correspondingly large compared with the size of the rock blocks.

b) Length and diameter of the test section

The lenght of the water test section is represented by L on fig. 4. It is not possible to give an optimum length for every test. This length depends on the mean spacing b of the joints and on the scale of the studied flow phenomenon (flow through a slope or under a dam etc.). The ideal test length is thus equal to the dimension of the network used by the numerical analysis for the flow, but this condition cannot always be attained in practice.

A test in a gallery or a well ($\phi > 2$ m) is more representative, but more expensive, compared to tests in boreholes. If, for financial reasons, the number of large scale tests is limited, then the first approach is to carry out water tests in boreholes and subsequently to confirm the results by using a gallery or a well. Borehole water tests allow a statistical consideration because of the large number of tests which can be carried out.

c) Test pressure

The interpretation of the water test is rendered difficult by using high pressure, because secondary phenomena influence the test results (turbulence, fissure deformation etc.). Particularly for radial flow, as in a water test, the very high gradient near the borehole causes turbulent flow. In addition the 10 kg/cm^2 pressure used by Lugeon test produces a big deformation of the medium near the borehole. The plotting of "flow rate against hydraulic gradient or water pressure in the borehole" gives the **characteristic curve** of the water test. Figs. 20 and 21 show a typical test result in the field. Every principal effect is represented. After a short linear phenomenon (1), turbulence effects are noticed (2). This effect is quickly compensated by the influence of the opening of the joints through the high water pressure (3). After a certain limit (between 4 and 6 kg/cm^2 in practice) the influence of the joint deformation is predominant (4). In very deformable rock effects as demonstrated by (2) and (3) can disappear.

In practice it is very useful to carry out some tests for the entire range as in fig. 20 so as to get an indication of the inaccuracies involved. The hydraulic conductivity however, is obtained only from zone (1) or, if not possible, then from zone (2). From the turbulent conductivity it is easy to get the laminar conductivity as follows (see above section):

(18) $$K_{lam} = A(K_{turb})^2$$

The shape of the complete curve of fig. 20 gives an idea of the deformability of the medium. Fig. 21 gives some concrete examples.

The amount of pressure which is to be applied during the test must be obtained from fig. 20. This test curve must be obtained initially before a large number of tests are carried out to establish the optimum working pressures for a rock type. It must be remembered that 1 Lugeon corresponds to 1 litre per metre per minute at 10 kg/cm² excess pressure but it can also be defined as 1 litre per metre per 10 minutes at 1 kg/cm² excess pressure.

d) Measured values

According to the above section the linear flow rate q/b or Q/L and two points on the piezometric line ϕ (r) lead to an interpretation of the test. The first point is given by a measurement of the water pressure in the borehole itself; the second must be chosen, if possible, in the vicinity of the borehole (see fig. 22). For this reason boreholes in the field should be drilled in two's, e.g.: 1–2 and 3–4 on fig. 22.

Very often in practice, financial considerations limit the number of boreholes that can be drilled. In such a case the radius of influence R is used as second point (R is the point where initial groundwater conditions do not change). The radius of influence is given by an empirical equation according to Sichardt (see Castany 1963):

$$R = 3000 \ (\phi_{ro} - \phi_R) \sqrt{K}$$

In this way it is always possible to approach the required hydraulic conductivity. Errors in the evaluation of the radius of influence do not affect test results very much. The terms Log R/r_o , $1/r_o - 1/R$ or $1/r^2 - 1/R^2$ from equations (11, 13, 14) do not depend very much on R because $R \gg r_o$. However, the best way to interpret the water test is to measure the piezometric head at two different points.

e) Type of water test

There is an essential difference in results between injecting in or pumping out. A more representative test is to pump out. But this is not always possible if, for example, the depth of the free surface in the borehole below pump is greater than 6 m. At the depths greater than this it is necessary to put the pump in the hole and the cost becomes formidable compared with the usual injection of water from ground level in a small borehole (e.g. NX).

3.5 Conclusion

To solve most practical problems in rock hydraulics it is always necessary to have information on *in situ* hydraulic parameters. All the hydraulic parameters can be represented by the concept of "directional hydraulic conductivities". These conductivities, which are to be used in mathematical or physical models, can be obtained from the field using the water test.

If the number of the hydraulically principal joint systems is one, two or three, then each joint system can be tested separately. In this way it is easy to describe the anisotropic behaviour of a jointed medium. In interpreting the water test results both the geometry and the regime in which the test is carried out must be considered. The working pressure (e.g. 10 kg/cm^2) in conventional tests is often sufficient magnitude to place the test conditions out of linear regime.

This method is not applicable if the number of hydraulically principal joint systems is larger than three or if the jointing is irregular. In such cases the permeability tensor of the medium can be obtained with large scale tests, assuming the medium as a continuum.

4 CONTINUOUS OR DISCONTINUOUS MEDIUM

Before starting on a study of the flow in a fissured medium, it is essential to determine whether the problem is to be considered as being continous or discontinuous. There is no general rule, and this notion only depends on the **relative scale** of the phenomenon studied and of the modulus of jointing characterized, for instance, by the mean distance between single fractures. This question of relative scale is outlined in fig. 23, which shows the same hydraulic problem, but for four different media.

It will be correct to consider a fissured medium as being continuous if the dimension of individual blocks is negligible as compared to the phenomenon considered (Case 2, Fig. 23) that is, if one can approximately count, say, 10,000 fissures in any plane section. On the other hand, if the number of fissures is between, for instance, 100 to 1000, the hypothesis of a discontinuous medium is necessary (Case 3) and finally if, in a given section, the number of fissures is less than 10, each fissure will have to be **individualized** in the mathematical or physical model used (Case 4). The number of fissures given here is subjective; in fact, the hypothesis to be chosen will have to be very carefully analysed for each given problem.

5. THREE–DIMENSIONAL DISTRIBUTION
OF THE HYDRAULIC POTENTIAL

5.1 Introduction

In this study, we shall not consider the problem of continuous media. It has been investigated by a number of research workers, either by numerical analysis or by electrical analogy. If a problem in rock hydraulics can, because of very close fissuring, be treated by the methods relevant to continuous porous media (for example, Case 2 fig. 23), we will only give the mathematical method for calculating the anisotropic permeability tensor from the hydraulic characteristics of the different systems of fissures.

A number of research workers have already studied the problem of three-dimensional flow within fissured media: Serafim (1965, '68), del Campo (1965), Romm (1966) Snow (1965, '67). They bring the problem down to a **tensorial** representation of the hydraulic properties of rock masses, traversed by three mutually-perpendicular systems of parallel fissures. Because of the tensorial notation, it is explicitly admitted that we are dealing with a **continuous medium**. However, this hypothesis is only very rarely verified in fissured rocks (see Paragraph 4).

In contrast to this approach, the methods developed in the present work, through the concept of "directional hydraulic conductivity", take into account the **discontinuous character** of the fissured rock masses, their **heterogeneity** in a given field and the **completely arbitrary orientation** of the network of fissures. Different methods are suggested to deal with the nature of the fracturation, starting from the simplest case (a rock mass with a single system of conductive fractures) to the complex case of $\underline{n}$ arbitrarily-oriented sets of fractures.

The methodology we have adopted is based on a fundamental property of the most general types of flow within a fissure. It has been shown that, in the steady state, the flow of water in a fissure, whatever its orientation in space, follows the potential theory when one uses as a velocity potential $k_f \phi = k_f (Z + p/\gamma_w)$, which is related to the hydraulic potential ϕ. This property also extends to a system of plane parallel fissures when the hydraulic conductivity k_f of a single fissure is replaced by the directional hydraulic conductivity K of the system of fissures.

From this important result, it becomes possible to apply the numerous methods of potential flow theory to each individual fissure or to a series of parallel fissures in a rock mass, as a whole. The problem in space, as a whole, is thus broken down into series of two dimensional problems. In each fissure, or system of parallel fissures, the domain of the hydraulic potential obeys Laplace's equation (harmonic potential). For each individual two-dimensional problem, there are numerous methods of solution: mathematical analysis (conformal mapping), numerical methods (among others, relaxation methods), graphical methods, and also the electric analogy method. As references, we shall quote the works of Dachler (1936), Polubarinova-Kochina (1962), Castany (1963), Schneebeli (1966), Irmay (1968), Bear (1972), to cite but a few of the most important ones.

It is therefore our intention to find, with the help of mathematical or physical models, a function $\phi\,(x,y,z)$ which verifies the equation $\Delta\phi = 0$ in each individual fissure within a domain D, knowing the directional hydraulic conductivities at the scale of the lattice of the model used, and admitting that function ϕ, or its derivatives taken perpendicular to the boundary, assume given values along the limit of the domain D (flux or potential conditions). Four cases will be considered, depending on the nature of fracturing in the medium.

5.2 Rock mass with one system of conductive fractures

Fracturing in the rock mass has the following characteristics:

a) A single system K_1 of main conductive fractures (plain or corrugated) and secondary fissures, in the hydraulic point of view of arbitrary orientation (fig. 24).

b) Two systems K_1 and K_2 of main fissures distributed in different domains and intersecting in a limited area. Systems such as K_1 and K_2 are said to be sequent. Secondary fissures may also exist. This case has been considered by Louis, 1967-69.

Fissures are said to be secondary if their contribution to the hydraulic potential in the rock mass is very small or nil.

In case (a), the flows in the different fissures of K_1 are quite independent. Each plane of fissuring will therefore be considered individually. In such a plane, the network of streamlines and the equipotential lines can be constructed by the usual methods of two-dimensional potential flow. Figure 24 (b) sketches the streamlines in any single fissure plane K_1 of figure 24 (a). Within the plane of the fissure, axis OZ is an ascending axis directed along the slope. For non-vertical fissures, it is, of course, necessary to multiply values in the vertical direction by sin θ (θ being the slope of fissure K_1 considered in figure 24). One can proceed similarly for all the fissures K_1 of the rock mass. By grouping all the results obtained in three-dimensional space, the total field of the hydraulic potential in the entire rock mass is obtained.

5.3 Rock mass with two systems of conductive sub-vertical fractures

In this paragraph, we consider a rock mass with two approximately vertical systems of conductive fissures, while the third network, horizontal in this case, is taken to be secondary. Figure 25 sketches the data of this problem.

This case is often met with in practice (e.g. the abutments of the Vouglans Dam). The systems of vertical fissures are, for instance, joints traversing the sedimentary layers, the horizontal network being made up of bedding planes which have become secondary (hydraulically speaking) by a clay filling.

Flows in the vertical networks are no longer independent. The determination of the hydraulic potential distribution must therefore take into account the mutual action of the fissures. The problem can be solved by a numerical method taking into account this reciprocal action; mathematically this is expressed by the continuity equation obtained by setting the two flows to be equal at the intersection of two individual fissures.

To draw up the model representing the rock mass, it is interesting to consider two cases:

1) When the **vertical dimension** of the domain under study is **negligible as compared to the longitudinal dimension**, the problem may then be reduced to one problem in two dimensions, if one takes as unknowns the **mean values** of the hydraulic potential ϕ_i along the vertical intersections N_i of two individual fissures K_1 and K_2. This is the case when the rock mass is isolated between two impermeable clay-banks (confined flow), or when the mass is not very high and is on an impervious sub-layer (free surface flow, figure 26). Under these conditions, the vertical component of the hydraulic gradient $\vec{J} = \overrightarrow{grad}\,\phi$ is negligible, as compared to the horizontal component.

In order to obtain the ϕ_i equation used to solve the problem, it is sufficient to state the equation of continuity for each intersection N_i of fissures K_1 and K_2 expressed by the conservation of flow. This law, known as "Law of Intersections" is given on the algebraic relationship (for the intersection N_i, figure 26):

(19) $$\sum_j Q_{ij} = 0 \qquad i = 1, ..., n$$

Q_{ij} is the flow rate in the individual fissure connecting two consecutive intersections N_i and N_j (figure 26). A flow will be taken to be positive if it occurs towards the intersection, negative in the opposite case. The application of this law to all the intersections of fissures K_1 and K_2 in the given domain will yield a system of n equations in Q_{ij} equal to the number of intersections, that is, to the number of unknowns ϕ_i .

Since the equations thus obtained are independent, the system is, in theory, solvable. It is sufficient to re-write the equations in terms of ϕ_i , the mean hydraulic potential along the intersection N_i. By taking into account the values of the directional hydraulic conductivities measured *in situ* along a length corresponding to the lattice of the model, the equation (19) becomes

(20) $$\sum_j K_{ij} \frac{\phi_j - \phi_i}{\ell_{ij}} A_i A'_i \ell_{ik} = 0$$

which finally reduces to:

(21) $$\sum_j K_{ij} \frac{\ell_{ik}}{\ell_{ij}} (\phi_j - \phi_i) = 0$$

The problem can easily be solved on a computer by various iteration methods. Once ϕ_i , the hydraulic potential at a fissure intersection, is known, the problem may be considered as solved. It should be noted that in this case, the free surface flow is known after a single computer run.

2) If the **vertical dimension** of the given domain is about equal to or **greater than** the longitudinal dimensions, it becomes necessary to take as unknowns a number of the hydraulic potential along the vertical intersection of two fissures K_1 and K_2, this being equivalent to quantifying the flow along the vertical dimension. The method of computation is then very much akin to the method of **finite differences**, with the vertical lattice network distributed in space. With respect to the previous problem, the number of unknowns is multiplied by the number of horizontal layers considered. The method of solution on a computer is however identical to the previous one.

If the phenomenon under study also exhibits a free surface, it will be necessary to perform successive iterations to determine it, knowing that the final type of flow must satisfy the equation $\phi \geqslant z$ at all points of the flow area.

One may thus solve a case where water emerges as springs or seepage from a rock slope, for instance. The heights of seepage along the cracks must be determined with the help of an empirical law, this law having been verified by measurements on a model.

Flow phenomena in the neighbourhood of the intersection of two fissures have been studied experimentally. It was found that in the case of the problem sketched in figure 26 and when the gradients are weak (say, less than 0.5), the hydraulic potential along the intersection of two vertical fissures varied but little and could even be taken as constant.

Figure 27 clearly shows the discontinuous character of the flow in fissured media; along the intersection of two fractures (with a ratio of fracture openings equal to 2:1), there occurs an important seepage surface with a resulting ratio in the piezometric heads of 1:4.

5.4 Rock mass with three systems of conductive fractures

a) Perpendicular fractures

The solving of problems of flow within a medium containing three systems of arbitrarily-oriented fractures, is evidently more complex.

It is noteworthy that in nature the principal directions of fracturation are very often approximately orthogonal. Therefore, the hypothesis of the tri-orthogonality of fractures is usually reasonable. This is why this particular case will be studied in greater detail, after defining the elements allowing for the construction not only of a **mathematical model**, but also of two **physical models**, one **electric** and the other **hydraulic**.

Let us first consider the **numerical method**. The mathematical model is built on a tri-perpendicular mesh, as shown in figure 28, which represents an elementary block of rock, and the six nodes (corners of the blocks) adjacent to the central node C, referenced by three indices i, j, k or by a single one, according to the technique of programming. As before, the three directional hydraulic conductivities $K_{u, v, w}$ of the fracture or of the system of fractures are known (see chapter 3). The equation of continuity giving the solution of the problem is still obtained by writing that the sum of flows at node C is zero, which gives

$$q_e + \sum_{o..} (K_u + K_v) M_{oc} (\phi_o - \phi_c) = 0 \qquad (22)$$

$\sum$ for N, S, E, O, H, B nodes, and q_e stands for the external flow injected into (+)

or pumped from (−) C. M_{oc} is a coefficient which only depends on the length of the edges of the mesh. In the general case, its value is:

(23) $$M_{oc} = \frac{A}{OC} = \frac{(NC + SC)(HC + BC)}{4\ OC}$$

In the case of cubic mesh, (of parameter a), relationship (22) reduces to:

(24) $$q_e + \sum_{o..} a(K_u + K_v)(\phi_o - \phi_c) = 0$$

Formulated in this manner, the problem can be programmed for computer treatment with no difficulty. As in the preceding cases, the unknowns are the potentials at the nodes; each equation such as relation (22) thus contains at most seven unknowns.

In the **physical models** (electrical or hydraulic) it is sufficient to simulate the linear relation (22) by a simple physical phenomenon such as the potential drop in an electrical resistance or the loss of head in a circular pipe in which fluid is flowing.

The **linear elements** of the three-dimensional meshed network in the electrical or the hydraulic analogue can be obtained by the equivalence among the three fundamental magnitudes (for element OC for instance): the coefficient of the drop in piezometric head for the fractured medium $(K_u + K_v)M_{oc}$, the electrical conductance 1/R + the coefficient of drop in piezometric head of the circular pipe $C_{uv}\ M_{oc}$. These equivalences are illustrated in figure 28. In the hydraulic conductivity C_{uv} of the circular pipe is obtained from Pouseuille's Law. By relating the flow to cross-section A (figure 21), we obtain

(25) $$C_{uv} = \frac{\pi g\ d_{uv}^4}{128\ \nu\ A}$$

A two-dimensional analogue model with 594 resistances, adjustable from 0 to 10 KΩ for 352 nodes, has been constructed at Imperial College in London by Sharp (1070); it is completely automatic, and the results are given through a computer as equipotential lines.

An element of the meshed network for the hydraulic model is shown in figure 30. The edges of the mesh are the circular pipes and form the basic parallelepiped of the network. The nodes are made out of small perspex prisms (see detail 1, figure 20), and they hold the pipes together. These must be sufficiently rigid to enable the assembly to stand up by itself. It is therefore advisable to use hollow perspex pipes.

Each individual pipe of length ℓ_{uw} represents hydraulically the permeability of the rock mass in the direction parallel to the two fractures U and W, i.e. in the direction of their intersection. The fundamental point in this similitude is that the **same ratios** as those occurring in nature, concerning hydraulic conductivities parallel to the intersection of the fracture planes, must be retained.

In the hydraulic model, the distribution of the hydraulic potential is determined from measurements of piezometric heights taken, for instance, at each node of the model.

In underground hydraulics, the **determination of the free surface** constitutes a delicate problem. It can be made fully automatic on mathematical models; but this considerably increases computation time. On electrical analogue models it is unfortunately manual, and this constitutes a long and tedious step. This same problem is, naturally, quite easily solved on hydraulic models, since the free surface immediately becomes apparent in each case considered.

Utilizing hydraulic models therefore makes it possible to consider a great number of cases without introducing further complications. Thus, for instance, if one wanted to study the effect of a grout curtain or a drainage system, it would be sufficient to make slight modifications (such as the closing or opening of a few pipes) to see immediately the consequences of these modifications.

On the other hand, electrical models have the advantage of allowing for **automatic readings of potential at the nodes** through electronic systems. This only becomes possible in hydraulic models with the incorporation in the perspex prisms of pressure transducers in place of the water-column piezometric tubes. Such an electrical system would, of course, be quite expensive.

b) Arbitrarily-oriented fissures
The models suggested in the preceding paragraph are made up of **line elements**: each fracture plane, isotropic or anisotropic, is represented hydraulically by two main permeabilities which must be orthogonal (this being due to the fact that an elementary fracture is a **continuous medium** in two dimensions). In three dimensions, this type of representation is only possible if the fracture in different systems forms a right-angle.

In the case of media with a triple fracturation of arbitrary or random orientation, we are obliged to use models with **surface elements**. These models are of two types: with triangular elements or with parallelograms or rhombus-shaped elements. In these simulations the finite element technique is recommended.

Methodology is therefore relatively simple: as shown in fig. 31, one lays out, in each individual fracture in space, one or more elements (triangle, parallelogram or rhombus) and it will then be sufficient to write, as done previously, that the sum of the flows along an edge or within a mesh is algebraically zero. This type of iteration problem is easily solved with a computer.

5.5 Rock masses with more than three fracture systems

We now consider the case of a rock mass with $\underline{n}$ systems of parallel fractures $(n > 3)$. The study of three-dimensional flow through such a medium is evidently quite complex. One could represent the flows by placing within each elementary fracture a certain number of triangular elements and then applying the techniques of finite elements. This method would require exact knowledge of the geometry of the fracture network, which in the case under consideration is very complex, as more than three systems of fractures are involved and as, furthermore, it would mean a large number of elements.

The method of finite elements therefore proves here to be **difficult in practice.** It is possible to approach the problem in a different light providing that the medium can be considered as continuous. This assumption may be valid when more than three fracture systems are present. It will then be enough to determine the permeability tensor, and thus reduce the problem to a case of a porous anisotropic medium, for the study of which there are already a great number of techniques. In any case, knowing the principal directions of permeability $K_{1,2,3}$, it will always be possible to solve the problem completely, using the methods suggested in paragraph 5.4 (a). One merely replaces $(K_u + K_v)$ in relation (12) by one of the principal permeabilities $K_{1,2,3}$.

The problem is simply one of determining the permeability tensor. One therefore considers $\underline{n}$ systems of fractures, shown stereographically in figure 32, of hydraulic conductivities K_i (K_i is measured *in situ* or is determined from relationships (3), (7) or (8)). In a system of perpendicular axes x_i, y_i, z_i associated with the system of fissures K_i (and eventually with the principal directions of permeability in the system K_i) the permeability tensor — with oz_i perpendicular to the system, can be written:

$$\overline{\overline{K}}_i = \begin{vmatrix} k_i' & 0 & 0 \\ 0 & k_i'' & 0 \\ 0 & 0 & k_m \end{vmatrix} \qquad (26)$$

k_i' and k_i'' are the principal permeabilities of the fracture system K_i and k_m is the permeability of the rock matrix. Let P (i) be the transformation matrix

$$x_i , y_i , z_i \xrightarrow{\quad P\ (i)\quad} X,Y,Z$$

In the system of axes X, Y, Z (figure 32), the total tensor of permeability due to the presence of n systems of fractures K_i is therefore written as (property of linear transformations):

$$\overline{\overline{K}} = \sum_{i=1}^{n} P(i).\overline{\overline{K}}_i . P^{-\ell}(i) \qquad (27)$$

This method is quite general and is applicable whatever the number of fracture systems (1,2,3 ...n), on the express condition that the medium may be considered as continuous (see paragraph 4).

6. DRAINAGE IN FISSURED ROCK

6.1 Introduction
Development of mining techniques, as well as progress in methods of construction in rocks, have made it imperative for engineers to concentrate on improving rock masses. Apart from rock grouting and the anchoring of rock faces, drainage constitutes one of the most efficient processes used to improve the stability of a mass of rock and to obtain rational and economical operation of mining works or of above ground or underground construction activities.

It is well known that groundwater, flowing or at rest, greatly endangers the stability of rock masses. The basic aim of drainage is to eliminate or reduce the mechanical or phsyco-chemical action of water flowing within the rock.

We are here touching upon a specific aspect of the problems facing the rock-hydraulics engineer, namely the drawing up of rational projects for drainage in fissured rocks. For their calculations to be correct, they must first of all take into consideration the medium and the laws governing water circulation; a specific methodology for the solving of drainage problems must also be evolved. The factors guiding the choice of one or the other drainage systems must always take into account the mechanical aspect of the problem; it must be kept in mind that the purpose of drainage is to reduce, as much as possible, the undesirable mechanical effects of water in order to improve the rock stability. Faced with such needs, the rock-hydraulics engineer must, at the same time, be proficient in mechanics; all this points to a very specific field of knowledge, which has recently been named "hydrogeotechnics".

Having recalled the hydraulic characteristics of rocks and analyzed drainage phenomena in such media, we shall then review some general topics which need to be taken into consideration if a rational drainage network is to be evolved. The methodology suggested may well be illustrated by a concrete practical example: drainage of a slope.

6.2 Project of a drainage network

6.2.1 Preliminary survey

When planning a drainage system, it is, of course, imperative to make the usual preliminary geological survey with a detailed analysis of the structural aspects involved. A thorough topological survey of all the joints is essential to lay down a rational plan of pumping tests (chapter 3). This will help to map out hydraulic conductivities in the fields studied and thus to define the details of the model that is to represent the medium (chapter 5).

6.2.2 Drainage criterion

Before undertaking drainage calculations, it is necessary to make a proper choice of the drainage criterion which will enable the engineer to determine, from both a qualitative and a quantitative point of view, the relative merits of various solutions. Each given problem has several drainage solutions. Choice of the optimum solution is essential.

A drainage criterion depends on the type of drainage problem faced. It would be different for a slope, a dam or a gallery. In order to define the criterion, one either determines a **drainage area** at the limit of which the resultant of all mechanical actions of the water are calculated, or one determines **a line or a surface along which the pressures will have to be minimal.**

In the particular case of a slope, the area to be drained with a minimum number of drains corresponds to the critical sliding area. That area is generally unknown. To simplify the problem, the area to be drained has been bounded, inside the slope, by two segments (fig. 33) so as to make the limit thus defined coincide, as nearly as possible, with the critical sliding curves obtained by the usual classical methods of soil mechanics (fig. 34). Whatever the angle of the slope, the surface of the drainage area thus defined will be constant:

$$A = \frac{3 \ H^2}{4} \quad .$$

For a given drainage system and given flow conditions, the resultant of all forces due to the water flow along the drainage area limit will thus be calculated as well as the dimensionless coefficient f (f = $\gamma_w A/F$), characteristic of the action of ground water, alternatively the efficiency ℓ/f of the drainage network. It is this coefficient f, or its inverse, which will be taken as a **drainage criterion**. For perfect drainage, the coefficient of water action will be nil and the drainage efficiency infinite. In fact, f, or its inverse, will be a vector; it will be related to the direction of the resultant $\vec{F}$. In choosing the best solution, it may be interesting to take into consideration not only the magnitude of this vector, but also its direction; in problems concerning stability, the direction of the applied force may indeed be of the utmost importance in the stability of the system.

6.2.3 Theoretical calculations

It will be useful first to determine the flow network without drainage and then to work out the water action coefficient, f_o. Then one can study, for the particular drainage system considered, the different flow networks by determining the potential grid and each coefficient $f_{1,2,...\underline{n}}$ to go with each solution. It will thus be easy, through the f coefficient, to find the optimum solution.

Generally, for a given problem, it is best to begin by determining the **optimum direction** θ_{opt} of the drainage system for a given length chosen beforehand and then by the same procedure to find the **optimum length** L_{opt} of the drains that corresponds to the fixed θ_{opt} direction of drains.

A study of the curves f (θ, L) first plotted with θ variable and L fixed, then with $\theta = \theta_{opt}$ and L_{opt} as shown in fig. 35 only qualitatively illustrates the case in point. Strictly speaking, a systematic study of the function f(θ, L) of the two independent variables would be needed in all the two dimensional space θ, L within the fixed limits of the problem at hand.

In most of the actual cases, drainage networks are made with parallel drillings, 5 to 10 cm in diameter. There is no criterion to help determine the distance d between drains; it is generally set between 2 and 10 metres.

The actual flow in a drained slope is three-dimensional, hence extremely intricate to study. Two-dimensional studies therefore give only approximate solutions. In cross-section, the drainage is only complete in the planes of the drains. Between drains, the drainage is limited. In consequence, the real depth of drains will have to be set at a value L greater than the theoretical L_{opt} value computed (fig. 36) L_{opt} is the **effective length of drains** whose real length is $L . L_{opt}$ is also the depht of the continuous drainage trench equivalent to the parallel drain network. One may thus assume that the relationship between the actual length of the cylindrical drains and the width of the area drained — corresponding to L_{opt} in the theoretical calculations — is of the form:

$$L = \alpha . L_{opt} ,$$

where the coefficient α is a function of distance d between the drains (for practical purposes, one selects $\alpha = 1.5$).

Strictly speaking, a study of the flow network in the plane of the drains would be necessary if one were to determine the true value of coefficient $\alpha(d)$.

6.3 Drainage of a slope
6.3.1 Description of the problem

Slope drainage problems are generally solved in practice by one of the three following techniques (fig. 25):

a) By drainage from the foot of the slope with a network of parallel cylindrical drains.

b) By natural drainage using a drainage gallery parallel to the surface of the slope.

c) By pumping from wells drilled from the top of the slope.

Solution (a) is certainly the most economical. It is thus most frequently adopted. Moreover, it is particularly suitable for a low slope (roughly ten meters high).

Solution (b), which is more costly, is on the other hand more efficient than solution (a) owing to the fact that the drainage gallery is parallel to the surface of the slope. Moreover, the actual width of the area drained is known. This solution can be justified only in the case of a large slope (with a height exceeding 100 m), for instance, in opencast mines or dam abutments or some slopes. The particular type of drainage has been analyzed by Sharp (1970) within the same study program as the present chapter.

Solution (c) must be adopted only where natural drainage (i.e. by gravity) is not possible, that is, when the foot of the slope is on a level lower than that of the natural surrounding ground. These conditions are frequently met in open-cast mining. The techniques used in such cases are very similar to those connected with ground water lowering. This solution is, of course, very costly, first of all because of the length and diameter of the drillings (for a depth exceeding 6 to 7 metres, pumps must be placed in the wells which calls for large diameters), and furthermore because of the energy used in pumping.

Whatever the solution adopted, the basic principle in the study of a drainage system remains the same. In each case, the best direction and area covered by this drainage system must be determined from the different flow nets using the same drainage criterion.

In this report, only solution (a) mentioned above has been taken into account in the case of a slope forming a 75° angle with the horizontal. The geometric and hydraulic data of the examples studied are shown in fig. 38. Boundary conditions are determined by the slope itself, by the surface with a constant potential H_w = 1.5H situated inside the mass, at a distance 3H from the foot of the slope, and finally by two impermeable surfaces or flow-lines for the lowest and highest boundaries in the domain under consideration. Directional hydraulic conductivities vary from case to case as regards both direction and relative value. In practice, the direction of the elementary hydraulic conductivities and the degree of anisotropy — defined as being the ratio of hydraulic conductivites — are to be measured *in situ*. Recent publications give all necessary theoretical and practical details to carry out these measurements (Louis and Maini, 1970; Louis, 1970).

In each case considered, boundary conditions are identical (main hydraulic gradient 0.5); only the hydraulic characteristics of the medium vary (such as joint direction and degree of anisotropy). In this study, two main fissure directions K_1 (of maximum hydraulic conductivity), have been taken into account: in the first case, fissures K_1 have a 30°, angle of dip upstream; in the second, the dip of K_1 is 15° downstream. The degrees of anisotropy were taken to be equal to 2 and infinity. So as not to complicate this report, only results concerning one example will be explained in detail below.

6.3.2 Theoretical results

The example chosen to illustrate the methodology detailed in paragraph 6.2 is outlined in figure 38a. The slope angle is 75°, the system of main fractures has a 30° dip upstream, while the secondary fissures are parallel to the slope surface. Results given in this report concern a degree of hydraulic anisotropy equal to 2. In accordance with the preceding discussion, the ideal direction of the drainage system has been determined first, drain lengths being dealt with subsequently.

a) Optimal direction of drainage network

Flow nets (equipotential lines, free surface) have been worked out by the now well-known methods of hydraulics of jointed media (Louis, 1970; Sharp, 1970, etc.), first in the undrained slope and for draining directions at -23°, +30° and +68°, for a fixed drain-length at 0.45H. With results concerning the drainless slope with -75° and +180° orientation, the first characteristics curve is given by 5 points (fig. 35a). Flow nets for each case considered are shown in figure 39, while figure 40 gives the efficiency curve of the corresponding drainage. From the general behaviour of the curve, it is convenient to determine within a few degrees the value of θ_{opt} at which curve $f(\theta)$ is minimum. In the present case, the value $\theta_{opt} = 15°$ was found.

b) Optimal length of drain

Following the same technique, flow nets have been studied for a fixed drain direction ($\theta = \theta_{opt} = 15°$) and various drain lengths, L = 0.20H; 0.40H; 0.50H; 0.60H. Results have been grouped in figure 41. The draining curve $f(L)$, shown in figure 42, has been worked out from 6 points, knowing that it is tangent to the straight line $f = 0$. A close examination of this curve makes it possible to conclude that the optimal drain length is situated in this specific example considered, between 0.5 and 0.6H.

6.4 Conclusion

Planning a drainage network proves in practice to be a complex problem. It is imperative, in order to solve it — and this essential for safety — to carry out a systematic hydrogeotechnical study of the operations, namely to ensure the stability of a slope by eliminating the disastrous effects of ground water.

Very often, analyses of stability and mechanical problems are carefully dealt with while the working out of network drainage characteristics are not studied in detail. It is advisable to harmonize the relative importance given to each phase of study in the framework of an overall program bringing in geology, mechanics and hydraulics simultaneously.

Practical experience shows that drainage networks are very often ineffectual. Their efficiency could be much increased and their cost reduced simply by studying drainability characteristics and by taking into account the structure of the medium (lithology, fissuring, discontinuity), simple hydraulic conductivities and the geometry of the medium.

Much progress still remains to be made in the fields of drainage. Theoretically, actual flows are often three-dimensional and transient, and their study therefore very intricate. Moreover, the geological aspects of the problem are important: certain masses of rock, where flow proceeds through small channels, for instance, are very difficult to drain by commonly used techniques.

It would appear that the use of explosives in drains could solve certain problems by efficiently increasing the range of action. Many aspects could lead to extremely interesting research projects.

7. PRACTICAL EXAMPLES

7.1 Water flow in dam abutments

We shall endeavour to determine the flow net (distribution of the hydraulic potential, free surface and seepage surface) in a fissured rock mass constituting the foundation of a dam abutment (the problem is sketched out in figure 25). The rock mass is taken to be intersected by two sets of fractures K_1 and K_2 (joints) which break up the sedimentary strata. The stratification boundaries K_3 are taken to be of secondary hydraulic importance, since the presence of filling impedes the free circulation of water. The grout curtain is situated in the plane of the upstream facing of the dam (fig. 43).

The results of the computation are grouped on figure 44 which represents a perspective view of the dam. This representation makes it possible to give an overall view of the free surface of the fractures of the rock mass.

7.2 Flow conditions at the Grand Maison damsite

As part of the study plan for the Grand Maison dam in the French Alps, the "Electricité de France" offered the opportunity for undertaking a highly developed analysis of flow phenomena within the foundation mass of the damsite. This analysis, aimed at understanding the hydromechanical behaviour of the dam and its abutments, required both by the foundation force exerted by the arch and the various water actions within the rock mass. A study of the dam's stability formed the last stage of this complex programme which included geological, structural, hydraulic and mechanical studies.

This chapter only deals with the hydraulic treatment which from the start relies greatly on the structural data of the site. The complexity of the water flow phenomena within the fractured rock required the use of a new, and indeed original, methodology, as much for the theoretical approach in the treatment of structural data, as for the testing techniques both in the laboratory and *in situ*.

7.2.2 The damsite and the construction

The damsite is situated in the Eau d'Olle Valley between the cristalline massives of Grandes Rousses and Belledonne, north-east of Grenoble (France). The dam, an arch about 200 m high will rest on a rock mass composed of very compact, fractured amphibolites and gneiss. The transition between the ribbon gneiss and the amphibolites is progressive and does not constitute any hydraulic discontinuity. The schistosity is marked but closed. The possibility of water circulation is determined essentially by the fractures.

This rock mass, as indeed do most rock masses, constitutes a very complex geometrical medium as a result of the fracturing. The fractures, irrespective of their origin, play a preponderant role in the hydraulic system. The medium, which has a negligible permeability of matrix compared with that of the fractures, is characterized by discontinuities, distributed in an anisotropic and heterogeneous manner. It is for this reason that the structural study of the site has a particular importance.

The dam which will rise to the 1700 m contour will have its foundations in unweathered rock, with their base at level 1940. The valley bed is covered by about 50 m of recent alluvia and scree which have a strong permeability compared with that of the bed-rock. Because of this contrast in permeability, the hydraulic study has been carried out without taking into consideration the presence of the alluvia and lateral scree found on the valley slopes.

7.2.3 Structural study

The whole hydrogeotechnical study of the Grand Maison damsite is essentially based on the structural characteristics of the environment. Thus, the hypothesis concerning the hydraulic parameters of the foundation mass of the dam, the nature of the mathematical model used to simulate the water flows and, finally, the kind of hydraulic tests carried out *in situ*, depend very greatly on the interpretation of the surveys of fracture development made on the site.

The preliminary hydraulic study (Louis 1970) was carried out from existing structural and geological data. During these reconnaissance surveys, 7,500 structural elements were noted on the site. From these preliminary studies it was possible to characterize the structure by five families of fractures, two of which may be considered as secondary from the hydraulic point of view. The principal families, F1,2,3 are all subvertical, their direction remaining within an angle of 60°-70° of the bisector plane, approximately parallel to the dam on the right abutment (fig. 45).

Certain conclusions may be drawn from this first study:

1) There is a very marked anisotropy of the hydraulic conductivity within the horizontal plane, the strong conductivity being parallel to the right abutment.

2) The vertical hydraulic conductivity, parallel to all the fractures will be the highest.

A very clear dissimetry in the hydraulic behaviour of the two abutments could therefore be forseen.

Because of the importance of the conclusions arrived at, it seemed necessary to verify the earlier result on the fracturing of the rock mass. This work was carried out by a new survey of the fractures within the zones directly affected by the water flow under the dam. The object of this exercise was very precise – to be able to describe, in the most objective way possible, the potential routes of water flow within the rock mass. It included two distinct phases: (see chapter 3.1).

a) A systematic survey in the left and right abutments and the sub-fluvial reconnaissance galleries and also in the excavations open at that time on the abutments (fig. 46), using a very detailed check list, which made it possible to describe each structural element observed in the most precise manner possible.

b) A statistical analysis of field data with a possible "hydraulic weighting", designed to reveal the hydraulic characteristics, representative of the site. To facilitate this task and to be able to carry out very varied statistical treatments, the calculations were made by computer.

From this complementary study it was possible to define the homogeneity of the structure of the site and also to analyse the validity of the hypothesis arising from the theoretical treatment of the hydraulic problem (the hierarchy between the fracture families, characteristics of the anisotropic tensor of the permeability, mathematical model, water tests to be forseen etc.).

In general, the weighting factors clarify the diagrams by discounting the secondary families which are sometimes numerically rich but not very important from a hydraulic point of view. Five families appear with the traditional treatment, whereas only three principal families (F1,2,3) are seen on the stereonets by using weighting factors.

Following this structural approach to the problem, a test of the anisotropic permeability tensor was undertaken by introducing various, absolute and relative hydraulic conductivities of the three principal fracture systems. In this way the most favourable orientations for the hydraulic tests to be carried out *in situ* could be chosen in order to determine the quantitative character of the anisotropy of the site.

7.2.4 Natural flow conditions

Before beginning the very elaborate study programs of the damsite, it seemed desirable to first make a piezometric study of the natural flow conditions. The aim of this exercise was on the one hand, to verify whether or not the rock mass had a homogeneous distribution of piezometric heads in the sub-fluvial zone beneath the aquifer, and, on the other hand, to obtain all the necessary elements for making a calibration of a mathematical model in a later phase. A simulation of flow conditions around the sub-fluvial gallery can in fact be achieved, given that all the elements, (conditions at the boundaries, permeability and piezometric head distribution, flow rates within the galleries) are known. Figure 47 gives a transverse cross-section showing the situation of the principal elements: the valley bed (1) at level 1550 m, the roof of the bed-rock (2) the sub-fluvial gallery (3) and the boreholes drilled from seven lateral excavations in the gallery.

The flow network between the valley bed with constant piezometric heads 1546 and the gallery at level 1485 m was analysed by the original technique presented in chapter 3.2.5., giving a continuous piezometric log of the boreholes.

The hydraulic examination of the site provided the following information (fig. 47):

– the fractured rock mass has a good hydraulic homogeneity;

– important fractures may be discovered by hydraulic means;

– stress relief is more marked on the right abutment than on the left;

– decompression is very little marked in the valley bed. It reaches no further than 5 m beneath the roof of the bed-rock.

7.2.5 Hydraulic characteristics of the site
a) Directional hydraulic conductivities

According to the conclusions of the initial geological study, later confirmed by the structural analysis, it seemed that the foundation mass of the Grand Maison dam is cut by three principal families of sub-vertical fractures (F_1 - N110E, F_2 — N135E, F_3 — N170E) which are not perpendicular (fig. 45). The definition of the hydraulic characteristics of the site and the simulation of the three dimensional flows could therefore only be made by two means (Louis 1970):

a) by a "Surface Elements" model coinciding with the fracture planes; simulation being achieved by the finite elements method (with plane elements distributed in space);

b) by a "Line Elements" model, using the concept of an equivalent medium to that of the real medium, solving the problem by the finite differences method.

For the present phase of investigation it was the latter technique that was chosen. It implies several hypotheses, in particular:

— to use the finite differences method we must admit that **the principal directions of the permeability of the medium are constant throughout the zone under consideration.** This supposes that at every point in the medium, the relationships of the directional hydraulic conductivities of each fracture family remain constant.

— **the use of tensorial writing must be admitted** for the determination of the principal permeability directions which are characteristics of the equivalent medium to the real medium. The "tensor of permeability" concept is only applicable to continuous media. This mathematical tool has only been used in the **preliminary phase** to define the characteristics of the line elements or the principal permeabilities of the equivalent medium. The simulation was then achieved with an essential discontinuous "Line Elements Model".

The total permeability tensor $\bar{\bar{K}}(X,Y,Z)$, a function of the point $M(X,Y,Z)$ is easily determined from the elementary hydraulic conductivities K, aK, bK, of the fracture families F_1, F_2 and F_3 and from the permeability of the rocky matrix k. In a system of axes x_1, y_1, z, linked to the fracture family F_1, the elementary tensor of permeability is written:

$$\bar{\bar{K}}_1 = \begin{vmatrix} K & 0 & 0 \\ 0 & k & 0 \\ 0 & 0 & K \end{vmatrix} \quad \text{oz vertical.}$$

The same is true for fracture families F_2 and F_3 by replacing K with aK and bK.

By expressing these elements in the same system of axes X,Y,Z (by changing the axes of the matrix P(i)) the total permeability tensor is then:

$$K(X,Y,Z) = \sum_{i=1}^{3} P(i) \; Ki \; P(i)^{-1}$$

It is written in the form:

$$\bar{\bar{K}}(X,Y,Z) = K(\sigma,t)\bar{\bar{F}}(a,b,\theta \; i)$$

The total permeability tensor $\bar{\bar{K}}(X,Y,Z)$ is thus expressed by a tensor F which fixes the principal permeability directions (linked to the relationships of hydraulic conductivities of each fracture family and to the structure of the site). The coefficient $K(\sigma,t)$ fixes, on the contrary, the absolute rate of permeability, which depends on the position of the point considered (by the intermediary of the depth of the overburdening) and the state of stress σ due to the dam.

In the example, $\bar{\bar{F}}$ (a,b,θi) is constant for the whole model, whilst the absolute rate of permeability was taken in the form k = k_o' exp (-At), a formula chosen after *in situ* and laboratory tests (see next paragraph). The principal permeabilities were therefore the following (fig. 45 and 46):

Horizontal directions K_1 =   1.6K N 124° E
K_2 =   0.17K N 34° E
Vertical direction K_2 =   1.75K.

b) Influence of the state of stress

The hydraulic characteristics of the fractured media are very sensitive to the state of stress imposed by the very weight of the rock mass or by exterior forces (tectonic or mechanical). It was an indispensable part of the study of the Grand Maison damsite to evaluate this influence, in order not only to determine the variation of the permeability according to the depth but also to be able to estimate the effects of the pressure exercised by the arch dam at a later date.

The influence of stress was introduced into the formula for the tensor of permeability by the intermediary of **the absolute rate of permeability** $K(\sigma, t)$ (see preceding paragraph). The object of the exercise was therefore to find out the law governing the variation of K according to a normal stress applied.

In the theoretical expression of the hydraulic conductivity of a family of parallel fractures, three terms vary simultaneously according to the state of stress; the free opening e, the relative roughness k/Dn and the degree of separation of the fracture K.

For a family of continuous fractures, the hydraulic conductivity is, in fact, expressed by (Louis 1970):

$$K_F = B \frac{Ke^3}{C}$$

with B the term independant of σ

$$C = 1 + 8.8(K/D_h^{1.5})$$

When the fractures are closed by a stress σ, K decreases from 1 to 0, e tends towards 0 and C increases from 1 to about 2. The laws for the variation of these parameters are **unknown**, they depend on the mechanical behaviour of each type of fracture. Only an experimental approach seems to be realistic.

For the Grand Maison site the analysis of the phenomenon was carried out *in situ* and in the laboratory on three different scales.

— *In situ* **water tests**

The influence of stress was analysed after a statistical treatment of the permeability tests made in the boreholes at varying dephts in the homogeneous fractured formations. The examination of the very numerous results showed that the empirical law that best translated the phenomenon was written:

$$K = K_o \, \mathrm{Exp} \, (-\alpha\sigma)$$
$$\text{with } \sigma \simeq \gamma t$$

K_o being the initial permeability (on the surface) γt the weight of the covering (t the depth). Figure 48 shows this law of variation of the permeability according to the depth for a borehole situated on the Grand Maison site. As we shall see, the coefficient $A = \alpha\gamma$ is variable according to the zone (right and left abutments, valley bed) over the whole site.

Laboratory Tests

Permeability tests under stress were carried out in the laboratory on two different scales (see fig. 18):

    — on a 30 x 25 cm fracture,

    — on rock cores (ϕ 4 cm) parallel to the discontinuities.

Some very varied samples were analysed in this way (Louis and Ricome 1974 , Louis 1974) ranging from those with a very marked fracturing to those which were compact and homogeneous.

The laboratory results showed that the exponential type of law obtained from the statistical treatment of the results of the *in situ* permeability tests, translated relatively well the phenomena of permeability variation according to the state of stress, at least within the stress variation zone, having an important role in the vicinity of a construction such as a dam. A detailed analysis of the results is given in the publications mentioned in the bibliography.

Taking into considerations the tests carried out *in situ* in the laboratory, it has been possible to draw up, by cross-sections, a map of the variations of the absolute rate of permeability within the rock mass (fig. 49). Also on this cross-section we see the variations of coefficient A which translates the sensitivity of the medium to a stress variation. (A varies from 7.8 to 3.4 10^{-3} m^{-1}). This distribution is only **provisional** and was chosen for the preliminary phase of the simulation. It will be refined and improved, taking into consideration the new hydraulic data available (e.g. after the very deep boreholes on the abutments and in the valley bed). Knowledge of the law of variation of the permeability rate will also make it possible to take into consideration the influence of the pressure of the dam on this distribution.

7.2.6 Flow simulation

a) **The mathematical model**

As stated earlier in paragraph 7.2.5, the simulation was carried out by a tri-dimensional, "Line Elements", mathematical model, using the finite differences method. The flows simulated are of free surface and steady state conditions. The mesh of the model is variable, the permeability is anisotropic, the principal directions of permeability and the degrees of anisotropy are fixed, whilst the rates vary according to the distribution of figure 49.

The mesh chosen for the first phase of calculation is represented by figure 50 which gives four horizontal cross-sections at different levels. The model includes ten horizontal layers, each of them comprising 21 x 21 elements of variable edge dimension. The total number of elements is therefore 4410.

The boundaries of the model are impermeable, the topographical surface is of given potential ($\phi = 1700$ m in the reservoir, $\phi = z$ downstream of the dam.

b) Results of the simulation

The aim of the simulation was, on the one hand, to obtain the distribution of the equipotentials, the shape of the free flow surface and the seepage surfaces, downstream of the dam, and, on the other hand, to estimate the leakage flow. Figure 50 gives some horizontal cross-sections through the flow network at the levels 1700 m (Maximum retained level) 1625 m, 1525 m (in the dam zone developing from 1490 to 1700 m) and 1475 in the valley bed. Ten analogous cross-sections were established.

Anisotropic effects were very marked in the calculation hypotheses. On the right abutment the free surface makes an angle of 10 to 20° with the corresponding level line, whilst on the left abutment, this angle is greater than 130° (cross-section at level 1625 m). This tendency is confirmed by the other cross-sections. The right abutment is therefore naturally badly drained; the piezometric head is very high there. The contrary is true of the left abutment. In the valley bed, the flow network is also very dissymetric. Longitudinal and transverse cross-sections of the flow network are given on figure 51, in the directions defined on figure 50a.

It seems that the gradients below the dam are very steep and also that the flow network remains quite complex (transverse cross-section) because of the particular topography of the site.

c) Consequences of the results

The analysis of the results provided by the model must be made remembering its limitations; the topography and the geometry of the dam are, in fact, imperfectly reflected, in that the mesh chosen are parallelepipedic and of large dimensions. The results obtained are therefore subject to caution for the direct vicinity of the dam (in a zone having an area of about one mesh). Moreover, possibilities of drainage or injection were not taken into consideration at this stage of the study.

A more detailed simulation could be undertaken by a secondary model, including a similar mesh number, but applied to a more restricted zone within the vicinity of the dam and taking as boundary conditions for the secondary model, the hydraulic results provided by the first model.

At the present stage of reconnaissance, the examination of the flow network leads to the following qualitative conclusions:

    — on the right abutment the flow force will be large; it should be drained. Leakage flows here are not very great.

    — the left abutment on the contrary, will be well drained, the flow pressures will be weak. However the flows below the abutments will be relatively abundant. According to the absolute value of the permeabilities, the left abutment may later have to be treated by grouting.

    — in the valley bed, the gradients below the dam seem to be very steep.

The importance of anisotropy is the essential factor in this study, as it conditions the whole of the flow network. This anisotropy was verified by *in situ* measurements.

7.2.7 Verification of the hypotheses by in situ tests

The marked hydraulic anisotropy of the Grand Maison damsite and its particular orientation, causing a total dissymetry in the flow networks have **very important consequences** in the behaviour of the dam. A serious verification of the hypotheses of calculation, and also of the preliminary results, therefore appears to be very necessary. This verification will naturally concern the evaluation of the anisotropy by *in situ* measurements using a technique particularly adapted for the hydraulic problem of the Grand Maison damsite, (see chapter 3.2).

The essential aim of the *in situ* hydraulic tests is to verify the characteristics of the tensor of the permeability using the triple hydraulic probe and point piezometers in boreholes (fig. 10b and 12).

The verification of the calculation hypotheses were planned for the two abutments and in the sub-fluvial gallery. Five lateral test excavations were made for this purpose. The emplacement of the boreholes in each excavation is given on figure 48. The injection boreholes, 50 m long, are directed parallel with the presumed orientations of the principal permeabilities. Five boreholes will serve for the estimation of maximum permeability whilst two others (perpendicular to the preceding ones) will provide information concerning the weakest permeability. Only the tests on the left abutment have as yet been accomplished.

7.2.8 Conclusion

Some very interesting conclusions concerning the hydromechanical behaviour of the dam abutments have resulted from the hydraulic study forming part of the Grand Maison Dam project. It seems that, at the present stage of the work, the most delicate problem is not the theoretical study of water flows but the determination of the hydraulic parameters. The study cannot, as yet, be considered as concluded. The reconnaissance campaign must be continued with the greatest of care, directing the programmes so as to gain the maximum information concerning the directional hydraulic conductivity values. The structure of the medium seems to be already well defined. Complementary work should be carried out for the study of the influence of stress on the flow networks.

Such a study is enlightening as it makes possible a detailed analysis of the stability of the abutments, the final phase of the geotechnical study for this development. This phase will lead to a precise determination of the drainage network and the grout curtain, which in the case of the Grand Maison site will be dissymetric, with its particular situation. If required, more details on this project are given in the recent papers (Louis 1970, 1972, 1974, Louis and Pernot 1972 etc...).

7.3 Flow in a fissured rock slopes

The characteristics of the example now considered are analogous to those of the example in paragraph 7.1. The difference lies in that in a rock slope, the flow is more nearly perpendicular to the surface of the rock mass than in the abutment of a dam. The data for the problem are given in figure 52. The boundary conditions are made up of the equipotential surface $\phi = H$, by the free surface of the rock slope $\phi = Z$, and by the lateral impervious surfaces. In order to emphasize the three-dimensional character of the examples considered the first boundary condition $\phi = H$ has not been chosen parallel to horizontal lines of the slope. As before the fracturing consists of two sets of sub-vertical fractures K_1 and K_2 and of a secondary horizontal fissure.

The results of the hydraulic simulation are given in figure 53 for two assumed orientations of the fracture set K_2. The distributions of the hydraulic potential are certainly three-dimensional and the seepage heights are not constant along the bottom of the slope. These simple examples prove again that, for given boundary conditions, the flow net depends strongly on the structure of the medium.

7.4 Hydraulic problems connected with dam projects

The two practical examples considered, which are studied in detail in other papers (Louis and Wittke 1971 and Wittke and Louis 1969), are very similar. Both are connected with two large hydroelectrical projects: one in Formosa (Tachien Project) and the second in West Germany (Bigge Dam Project). In each case the topographical situation is identical: upstream, the reservoir is connected through a very narrow rock ridge, with a valley which is parallel to the axis of the river. On the one hand, the leakage flow arising from this situation could have been very high, and on the other hand, because of the possibility of considerable hydraulic gradients, the stability of the rock mass was in danger.

7.4.1 Pitan Ridge (Tachien project)

The dam on the Tachia river is designed to retain water to a height of 1400 to 1420 metres. Immediately below the dam the river swings around and the valley extends in a direction parallel to the dam axis. The Pitan ridge thus constitutes a narrow separation between the reservoir (level 1420) and the valley below, at level 1230 (figure 54).

Beneath the superficial soils there is a jointed rock mass (made up or alternating quartzite and schistose strata). The fracturing is characterized by two perpendicular systems of parallel fractures.

The study of the flow within the rock mass has been performed with hydraulic models made up of PVC pipes (figure 55). In order to be sure that the study included the actual case, two different hypotheses of relative permeabilities were considered (cases a and b, figure 55). The advantage of the technique used was primarily to allow a low cost study of a large number of cases, as shown in figure 56 (different heights of the reservoir, different relative permeabilities, different drainage systems). An example of the results obtained, showing the flow in the rock mass without drainage, is given in figure 57. The study showed that the flow forces within the undrained rock mass are very large, and therefore that a network of drains was necessary to ensure stability.

7.4.2 Kraghammer Sattel (Bigge dam)

The Biggetalsperre hydro-electric complex is in the Sauerland, which is part of the schist zone of the West German Rhineland. The rock pass known as the Kraghammer Sattel separates, over an area of 300 m, the Bigge Valley reservoir from the Ihnetal, a parallel valley upstream of their confluence (figures 57 and 58).

Contrary to the preceding case, the level of the Kraghammer Sattel was below that planned for the reservoir. The rock mass, weathered at the surface and therefore quite permeable, was first to be given an increased height, then waterproofed to decrease losses. The stability of the downstream spillway was equally a problem because of water seepage.

A geological hydrogeotechnical survey of the rock pass was made with the help of the drilling, with water-pumping tests, of deep drill holes at the apex of the rock mass and of a longitudinal gallery. On the basis of the survey conclusions it was decided to install concrete piling in the stress relieved zone (which has a well-defined lower boundary, as seen during the water-pumping tests, figure 59), and also a grout curtain within the heart of the rock mass to a depth corresponding to the zones with very low permeability (as revealed by the water-pumping tests).

From the hydraulic point of view it was possible to distinguish, within the rock mass, two zones with clearly different characteristics (figure 59). The water-pumping tests and the statistical survey of fractures showed that flows within zone 1 occurred essentially through the bedding planes, which are quite open, also through sub-vertical transverse joints. As in the case of the Pitan Ridge, continuous bedding planes that are oriented parallel to the longitudinal axis of the rock ridge are prone to open up cavities or voids and thus to have a hydraulic conductivity markedly greater than that of the cross joints which are discontinuous; this fact was verified *in situ*. Because of this, a special hydraulic model was used to study flow phenomena in this domain (figure 60); the bedding planes themselves form equipotential surfaces. Zone II, a narrow faulted domain, is perpendicular to the longitudinal axis of the rock ridge and constitutes a large joint (with isotropic permeability) between the reservoir and the Ihne Valley. Flow phenomena were studied in zones I and II and various different boundary conditions were considered. Figure 61 shows, as an example, the flow networks in zones I and II for one of the case considered.

Along with the theoretical studies, a number of piezometric measurements were performed *in situ* in boreholes at the foot and in the heart of the rock mass. These measurements taken *in situ* proved to be in complete agreement with the hypotheses concerning zones I and II, and showed a gratifying concordance with the results of laboratory studies.

These two hydrogeotechnical studies were performed at the "Theodor Rehblock Flussbay laboratorium", of the University of Karlsruhe, in close collaboration with Professor L. Müller for the Pitan Ridge, and with the

"**Ruhrtalsperrenverein**" for the Kraghammer Sattel.

8. CONCLUSION

The recent development of computer techniques, particularly in the filed of scientific calculation, has greatly contributed to the progress achieved in these last few years in the hydraulic of fractured media and in hydrogeotechnical studies. The power of computer has considerably increased, and it is now possible to obtain representative results through the use of mathematical models with a sufficient number of elements.

Physical models, (electric or hydraulic), can also give quite serviceable and practical solutions to certain problems which had so far seemed insoluble.

The validity of results thus obtained depends, whatever the method used, on care in applying the techniques and on careful *in situ* determination of the basic parameters, namely the geometry of fracture development and the hydraulic condcutivities on which the model will be based. It is here necessary to adopt a realistic view: in rock hydraulics, it is inconceivable that one could describe in detail the **geometry** of the fractures in a rock mass because of their random and therefore unpredictable characteristics. The methods suggested are nevertheless applicable, for the essential requirement of mathematical or physical models is a knowledge of the overall hydraulic effect scaled down to the mesh of the model (as measured *in situ* with the help of appropriate testing techniques) and not a detailed knowledge of the fracture network structure.

On the basis of information obtained in the field the model can of course, be subdivided into elements as small as necessary. Hence, critical zones with a high gradient and also heterogeneous zones may be subdivided into a large number of elements, while other zones, less important hydrodynamically speaking and, more homogeneous, may be represented by an expanded mesh. The choice also depends on the precision required in final results. With this methodology, anisotropy, discontinuity and heterogeneity in naturally fractured media may be taken into account.

Before concluding, it is necessary to emphasize the need to carry out piezometric control measurements *in situ*, along with all theoretical or experimental studies. It is only through this method that one can test a mathematical or physical model and draw conclusions on the applicability of theoretical results thus obtained.

ACKNOWLEDGEMENTS

I wish to express my thanks to the Technical University of Karlsruhe in Germany, the "Rock Mechanics Center" at Imperial College, London and finally the B.R.G.M. in France, where I have been able to carry out research work on rock hydraulics.

May I also express my gratitude to the "Electricité de France" for their constant support, which made it possible for me to carry out *in situ* hydrogeotechnical studies at some major dam sites.

REFERENCES

[1]   Baker W.J. (1955): Flow in fissured formations. Proc. 4. World Petroleum Congress Rome Sect. II (p. 379-393).

[2]   Bear J. (1972): Dynamics of fluids in porous media. American Elsevier.

[3]   Bernaix J. (1967): Etudee géotechnique de la roche de Malpasset. Ed. Dunod, Paris. Thèse de Doct. (215 p.).

[4]   Bernaix J., Duffaut P. (1970): Auscultation piézométrique. 10. Congres des Grands Barrages, (1963) Montreal, Rep. 49. Q 38 (p. 935-960).

[5]   Castany G. (1963): Traité pratique des eaux souterraines. Dunod Ed., Paris (657 p.).

[6]   Dachler R. (1936): Grundwasserströmung. Springer Verlag. Wien (140 p.).

[7]   Duffaut P., Louis Cl. (1972): L'eau souterraine et l'équilibre des pentes naturelles, Bull. B.R.G.M. (2) III n° 4.

[8]   Dupuy M., Lefebvre du Prey E. (1968): L'anisotropie d'écoulement en milieux poreux présentant des intercalations horizontales discontinues. 3e Colloque de l'A.R.T.P.F., Pau Sept. 1968; Ed. Technip, Paris (p. 643-676).

[9]   Franciss F.O. (1970): Contribution à l'étude du mouvement de l'eau à travers les milieux fissurés. Thèse Doct. Jug. Université de Grenoble (1966-67).

[10]   Freeze R.A., Witherspoon P.A. (1968): Theoretical analysis of regional groundwater flow − 3 parts − Water Resources Research, 1966, V 2, n° 4 (p. 641-656) 1957, V. 3, n° 2 (p. 623-634), 1968, V. 4, n° 3 (p. 581-590).

[11] Gringarten A.C., Ramey H.J. (1971): Unsteady state pressure distributions created by a well with a single horizontal fracture, partial penetration or restricted entry. Soc. Petro. Engrs. of AIME.

[12] Gringarten A.C., Ramey H.J. (1972): Pressure analysis for fractured wells. Soc. Petrol. Engrs. 47th Annual Fall Meeting San Antonio. 8-11 Oct. 1972.

[13] Habib P., Sabarly F. (1953): Etude de la circulation de l'eau dans un sol perméable par analogie électrique à trois dimensions. 3e Congrès Intern, de Mécanique des Sols, II (p. 250-254).

[14] Hackeschmidt M.(1968): Der Aufbau eines räumlichen Widerstandsnetz-werkes mit Hilfe von elektrisch leitendem Papier. Freiberger Forschungshefte, Freiberg, Reihe Geophysique, C 214 (69 p.).

[15] Hills E.S. (1966): Element of structural geology Science Paperbacks — Methuen and Co., Ltd., London.

[16] Huard De La Marre P.(1955): Modèles analogiques à trois dimensions pour l'étude des écoulements de filtrations à surface libre. Comptes rendus Académie des Sciences, Paris, mai 1955, 240 (p. 2203-2205).

[17] Huard De La Marre P. (1958): Résolutions des problèmes d'infiltration à surface libre au moyen d'analogies électriques. Thèse Faculté des Sciences, Paris, juin 1956. Publ. Scient. et Thechn. du Min. de l'Air, n⁰ 340, (119 p.).

[18] Jouanna P. (1972): Effet des sollicitations mécaniques sur les écoulements dans certains milieux fissurés. Thèse Doct. es. Sc. Université de Montpellier.

[19] Kiraly L. (1971): Groundwater flow in heterogeneous anisotropic media. A simple two dimensional electric analog. Journal of Hydrology Vol. 12 n⁰ 3.

[20]   Lomize G.M. (1951): Flow in fissured rock — (russian) — Gosenergoizdet.

[21]   Louis Cl. (1967): A study of groundwater flow in jointed rock and its
        influence on the stability of rock masses. Ph. D. thesis University of
        Karlsruhe English translation, Imperial College Rock Mechanics.
        Research Report n° 10 London Sept. 1969 (90 p.) Univ. Microfilm.

[22]   Louis Cl. (1968): Etude des écoulements d'eau dans les roches fissurés et de
        leurs influences sur la stabilité des massifs rocheux. E.D.F., Bulletin
        de la Direction des Etudes et Recherches, Série A Nucléaire,
        Hydraulique. Thermique, n° 3 (p. 5-132).

[23]   Louis Cl. (1969): Sonde hydraulique triple pour déterminer les conductivités
        hydrauliques directionnelles des milieux poreux ou fissurés. Rapport
        inédit pour dépôt de brevet. Déc. 1969.

[24]   Louis Cl. (1970): Hydraulic triple probe to determine the directional
        hydraulic conductivities of porous or jointed media. Imp. Coll., Rock
        Mech. Res. Rep. G.B., n° D. 12.

[25]   Louis Cl. (1970): Ecoulement á trois dimensions dans les roches fissurées.
        Rev. Ind. Minér. Fr., n° spéc., 15 juillet 1970, pp. 73-93.

[26]   Louis Cl. Maini Y.N. (1970): Determination of in situ hydraulic parameters
        in jointed rock. Proc. 2nd Congr. int Soc. for Rock Mechanics, Belgr.
        Sept. 1970, T. 1, n° 1-32, (p. 235-245).

[27]   Louis Cl., Wittke W. (1971): Etude expérimentale des écoulements d'eau
        dans un massif fissuré, Tachien project, Formose. Geotechnique, 21,
        n° 1 (p. 29-42).

[28]   Louis Cl., Ricome B. (1972): Comportement hydromécanique de matériaux
        fissurés. Rapport B.R.G.M.

[29]   Louis Cl. (1972): Les drainages dans les roches fissurés. Bull. B.R.G.M. (2)
        III, 1.

[30]   Louis Cl. (1972): Les caractéristiques hydrauliques du massif de fondation
       du barrage de Grand Maison (Isère). Bull. B.R.G.M. 2 e série, Section
       III, n°4, (p. 13-137).

[31]   Louis Cl., Pernot M. (1972): Analyse tridimensionelle des écoulements dans
       le massif de fondation du barrage de Grand Maison. Symp. of the Int.
       Soc. for Rock Mechanics. Stuttgart T4-F1-16.

[32]   Louis Cl. (1974): Hydraulique des roches. Thèse Doct. es Sc. Université de
       Paris.

[33]   Maini Y.N. (1971): *In situ* hydraulic parameters in jointed rock. Their
       measurement and interpretation. Ph. D. Thesis Imperial College
       London Univ. University microfilms, Pen, Bucks (G.B.).

[34]   Mattauer M. (1973): Les déformations des matériaux de l'écorce terrestre
       Herrman Paris.

[35]   Müller L. (1963): Der Felsbau, Bd. I Enke Verlag, Stuttgart.

[36]   Opsal F.W. (1955): Analysis of two and three dimensional groundwater flow
       by electrical analogy. Trend Engineering University, Washington.

[37]   Peukert D. (1969): Betrag zur Lösung von Groundwasser-
       strömungsproblemen durch Spaltmodellversuche. Ph. D. Thesis
       University of Dresden.

[38]   Polubarinova-Kochina (1962): Theory of groundwater movement. New
       Jersey Princeton Uni. Press, 1962 (613 p.).

[39]   Prevosteau J.M. (1971): Analyse quantitative physique et géométrique de
       l'espace poreux des matériaux. Application aux roches grésseuses.
       Thèse de doctorat 3ème cycle. Université Orléan. 1971.

[40]   Price N.J. (1966): Fault and joint development in brittle and semibrittle
       rock Pergamon Press.

[41] Rayneau Cl. (1972): Contribution à l'étude des écoulements autour d'un forage en milieu fissuré. Thèse Ing. Doct. Université de Montpellier.

[42] Romm J.S. (1966): Drainage properties of fissured rock. Moscou, Izdat. "Nedra" 1966 (284 p.).

[43] Rushton K.R. Herbert R. (1970): Resistance network for three-dimensional unconfined groundwater problems with examples of deep well dewatering. Proc. Inst. of Civ. Eng. London., Vol. 45, March 1970 (p. 471-490).

[44] Schneebeli G. (1966): Hydraulique souterraine, Eyrolles Paris (362 p.).

[45] Serafim L.J., del Campo A. (1965): Interstitial pressures on rock foundations of dams. Proc. ASCE, SM, Sept. 1965 (p. 65-85).

[46] Serafim L.J. (1968): Influence of interstitial water on the behaviour of rock masses. Rock Mechanics in Engineering Practice. Ed. Stagg-Zinkiewicz. John Willey & Sons, London, New York, Sydney, (p. 55-97).

[47] Sharp J.C. (1970): Fluid flow through fissured media. Ph. D. Thesis imp. Coll. London 1970.

[48] Sharp J.C. (1970): Drainage characteristics of sub-surfaces galleries. 2nd Congress Int. Soc. of Rock Mechanics. Belgrad.

[49] Snow D.T. (1965): A parallel plate model of fractured permeable media. Ph. D. Thesis Univ. Calif. Berkeley 330 p. Univ. Microfilms.

[50] Snow D.T. (1967): Anisotropy of permeable fractured media. In hydrology and flow through media. Muskat Vol., R.J.M. De Wiest Ed., Academic Press, New York (86 p.).

[51] Starr M.R., Skipp B.O., Clarke D.A. (1969): Three-dimensional analogue used for relief well design in the Mangala Dam project, Géotechnique, Vol. 19, n° 1 (p. 87-100).

[52] Symposium of the International Society for Rock Mechanics and Association of Engineering Geology (1972): Percolation through fissured rock. (A collection of 36 papers) Stuttgart 18-19 Sept. 1972.

[53] Wilson C.R. and Witherspoon P.A. (1970): An investigation of laminar flow in fractured porous media. Publ. n$^{\underline{o}}$ 70-6, Universtity of California, Berkeley, Nov. 1970.

[54] Whiterspoon P.A., Noorishad J. and Maini Y.N. (1972): Investigation of fluid injection in fractured rock and effect on stress distribution. Geotechnical Engineering publication.

[55] Whiterspoon P.A., Neuman S.P. (1972): Finite element methods in hydrogeology, Seminar in B.R.G.M. Orléans. Rapport B.R.G.M. 72, SGN 374 AME (104 p.).

[56] Withum D. (1967): Elektronische Berechnung ebener und räumlicher Sicher und Grundwasserströmungen durch bleibig berandete, inhomogene anisotrope Medien. Mitteilungen des Institutes für Wasserwirtschaft und landwirtsch. Wasserbau, T.H. Hannover, Heft 10, 1967 (p. 180-233).

[57] Wittke W., Louis Cl. (1968): Modellversuche zur Durchströmung klüftiger Medien. Felsmechanik u. ingenieurgeol. Suppl. IV, (p. 52-78).

[58] Wittke W., Louis Cl. (1969): Untersuchungen zur Durchströmung des Kraghammer Sattels an der Biggetalsperre nach neuentwickelten Methoden der Felshydraulik. Mitt. der Versuchsanstalt für Wasserbau und Kulturtechnik, Theodor-Rehblock-Flubaulaboratorium, Universität Karlsruhe, Heft 157, (p. 131-210).

[59] Wittke W. (1970): Three dimensional percolation of regularly fissured rock. Proc. Symp. of open Pit Mines, Johannesb., Sept. 1970.

[60] Zinkiewicz O., Cheung Y.K. (1967): The finite element method in structural and continuum mechanics. McGraw Hill, London.

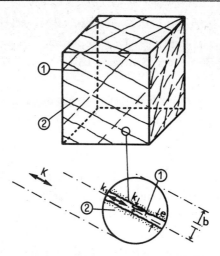

Fig. 1   Hydraulic Parameters of a rock mass
          (1) Fracture of the System Kj,
          (2) Rock Matrix,
          k_f, Hydraulic conductivity of a fracture,
          K, Hydraulic conductivity of the fracture set.

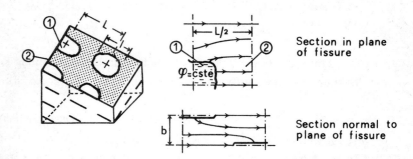

Fig. 2   Network of Discontinuous Fissures.
          (1) Open fissure,
          (2) Rock matrix.

1		2	3	4 Position			5 Pl.Obs.		6 Rock		7	8 Orient.		9	10	11	12	13	14	15	16	17	
p	n		dz	X	Y	L	Str.	Dip.	T_1	T_2	Str.	Str.	Dip.	Cont.	Th.	Fill.	Op.	112_0	Rel.	Sp.	Join	φ	Observations
				10	20	•		30				40		•		50					60		
						•								•									
						•								•									
						•								•									
						•								•									
						•								•									
						•								•									
						•								•									
						•								•									
						•								•									
						•								•									
				10	20	•		30				40		•		50					60		
						•								•									
						•								•									
						•								•									
						•								•									
						•								•									
						•								•									
						•								•									
						•								•									
						•								•									
						•								•									
						•								•									
						•								•									
						•								•									
				10	20	•		30				40		•		50					60		

6 Type of rock							
10		20		30		40	
11		21		31		41	
12		22		32		42	
13		23		33		43	
14		24		34		44	
15		25		35		45	

7 Structure	
1	
2	
3	
4	
5	
6	
7	
8	

9 Continuity Long. Rel.:	13 Water 0 No 1 Yes	14 Relaxation 0 No 1 Yes	15 Spacing (decimetres)	17 Friction

10 Thickness	
1	$<e_1$
	$e_1 =$
2	
	$e_2 =$
3	
	$e_3 =$
4	$>e_3$

11 Filling							
10		20		30		40	
11		21		31		41	
12		22		32		42	
13		23		33		43	

16 Joint continuity	
0	No
1	In the rock
2	Struc > 20°
3	Struc < 20°
4	Echelon

Fig. 3 Check list for systematic structural survey.

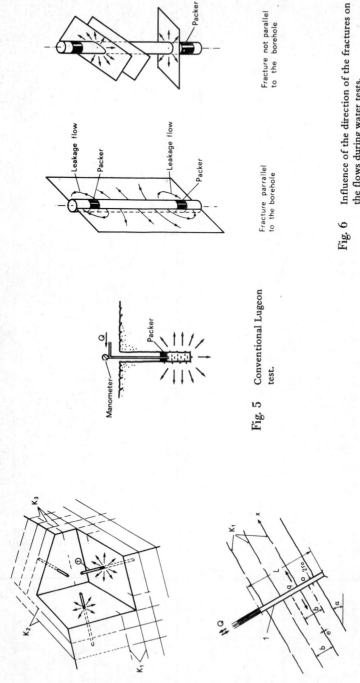

Fig. 5   Conventional Lugeon test.

Fig. 6   Influence of the direction of the fractures on the flows during water tests.

Fig. 4   Water test in rock mass with three systems of fractures
(1) Borehole parallel to K2, K3 for testing the fracture system K1.

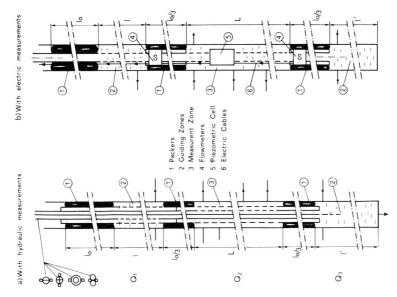

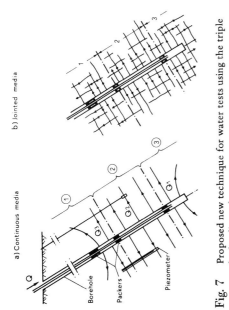

a) Continuous media
b) Jointed media
Q
Borehole
Packers
Piezometer
Q_3
Q_2
Q_1
1
2
3

Fig. 7 Proposed new technique for water tests using the triple hydraulic probe
(1) and (3) Three dimensional flow
(2) cylindrical flow

1
2
3

Fig. 8 Water test arrangement with four packers.

a) With hydraulic measurements
b) With electric measurements
Q_1
Q_2
Q_3
lo
l
lo/3
L
l'
1 Packers
2 Guiding Zones
3 Measurent Zone
4 Flowmeters
5 Piezometric Cell
6 Electric Cables
S

Fig. 9 Hydraulic triple probe for the determination of the directional hydraulic conductivities of porous or jointed media.

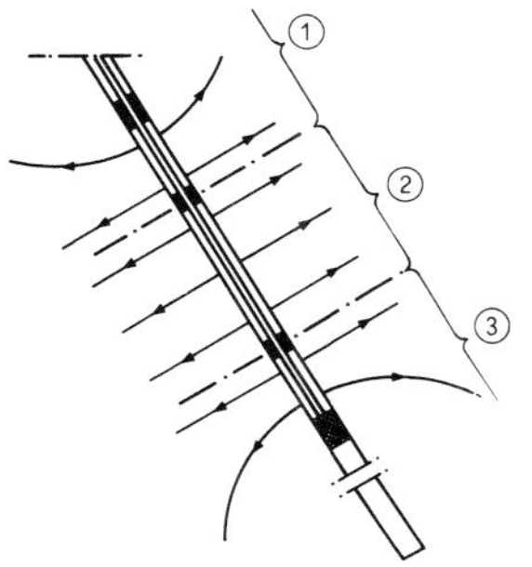

a) Investigations for the foundation of a large building (Diplodocus, Lille)

b) Measurement of hydraulic conductivities at Grand Maison dam-site (Isère)

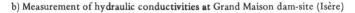

Fig. 10   Examples of testing with the triple hydraulic probe
(1) Triple hydraulic probe
(2) Lateral piezometers
(3) System to move the probe.

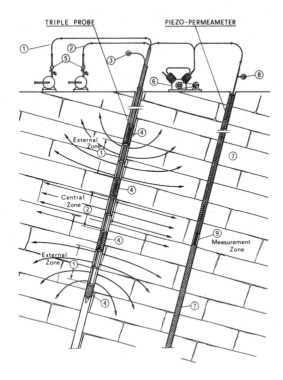

Fig. 11 Water test by using the triple hydraulic probe and the piezopermeameter

(1) Injection in the external cavities
(2) Injection in the central cavity
(3) Pressure in the central cavity
(4) Packers
(5) Flowmeters
(6) Compressor
(7) Generalised packer
(8) Pressure in the measurement cell.

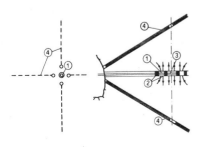

Fig. 12 Hydraulic test carried out at Grand Maison dam-site (see fig. 10b)

(1) Triple hydraulic probe
(2) Central injection cavity
(3) Piezopermeameter
(4) Measuring cell

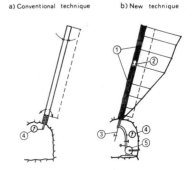

Fig. 13 Piezometric measurement in gallery
(1) Generalised packer
(2) Measurement cell
(3) Bleed valve
(4) Pressure gauge
(5) Pump

Fig. 15 Piezometry in borehole — Response
curve
(1) Air removal and water injection
(2) Air bleed
(3) Injection
(4) Pressure to be measured

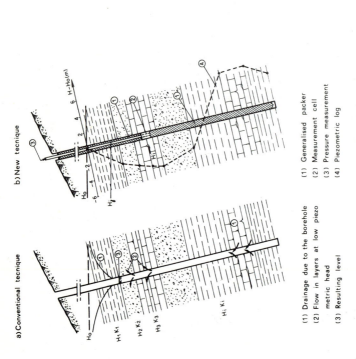

a) Conventional tecnique   b) New tecnique

(1) Drainage due to the borehole
(2) Flow in layers at low piezo
    metric head
(3) Resulting level

(1) Generalised packer
(2) Measurement cell
(3) Pressure measurement
(4) Piezometric log

Fig. 14 Piezometric measurements from the surface in multi-layered
aquifers
Ho Piezometric head of the free surface
Hi Piezometric head at any point M
Ip = Hi-Ho "Piezometric index"

a) Coupling the screen with the generalised pneumatic packer

b) Introduction of the piezopermeameter into a borehole

Fig. 16 Layout of piezopermeameter set-up.

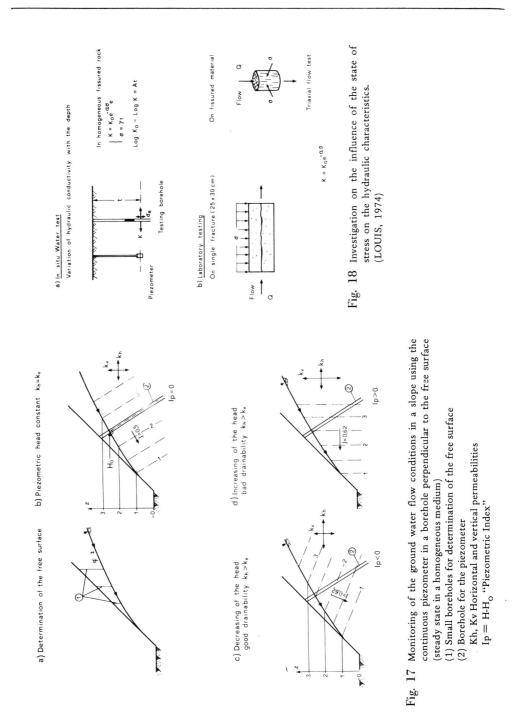

Fig. 18 Investigation on the influence of the state of stress on the hydraulic characteristics. (LOUIS, 1974)

Fig. 17 Monitoring of the ground water flow conditions in a slope using the continuous piezometer in a borehole perpendicular to the free surface (steady state in a homogeneous medium)

(1) Small boreholes for determination of the free surface

(2) Borehole for the piezometer

K_h, K_v Horizontal and vertical permeabilities

$I_p = H - H_o$ "Piezometric Index"

With a fracture system
in the testing area

With a single fracture
in the testing area

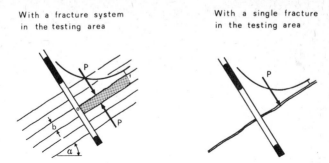

Fig. 19 Deformation of the medium during the water test.

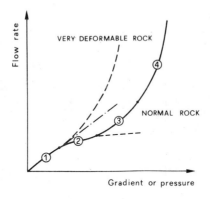

Flow rate

VERY DEFORMABLE ROCK

NORMAL ROCK

Gradient or pressure

Fig. 20 Typical results of field water test.
   (1) Laminar flow
   (2) Turbulence effect
   (3) Turbulence offset by fissure expansion
   (4) Predominance of fissure expansion effects.

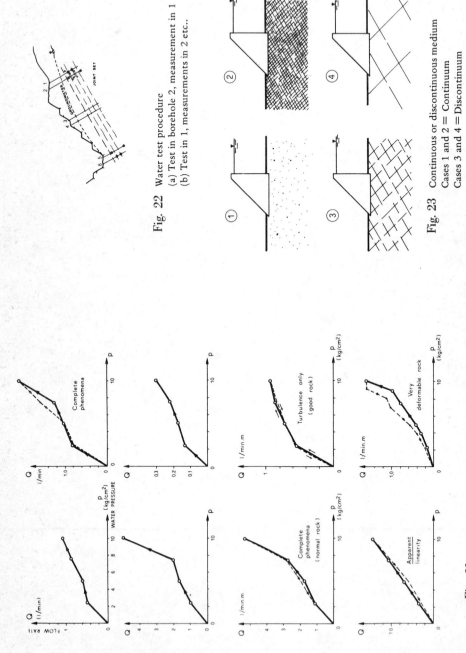

Fig. 22 Water test procedure
(a) Test in borehole 2, measurement in 1
(b) Test in 1, measurements in 2 etc..

Fig. 23 Continuous or discontinuous medium
Cases 1 and 2 = Continuum
Cases 3 and 4 = Discontinuum

Fig. 21 Practical results of in situ water tests.

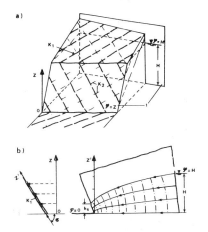

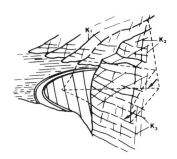

Fig. 25   Three-dimensional problem in rock hydraulics — Flow in a dam abutment $K_{1,2,3}$: Fissure surfaces.

Fig. 24   Rock slope with network of conductive fissures and flow network in a fissure of the rock mass.
(1) Boundary conditions.

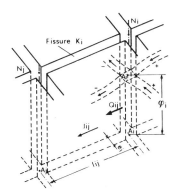

Fig. 26   Free surface flow in the neighborhood of the intersection of two vertical fissures.

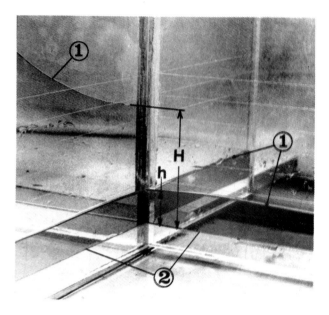

Fig. 27 Experimental study of the flows in the neighborhood of the intersection
of two fractures.
(1) Free surface
(2) Impermeable boundary
H, h: piezometric heights.

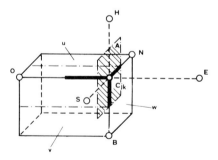

Fig. 28 Elements of the mathematical tri-orthogonal model.

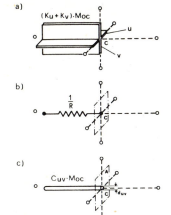

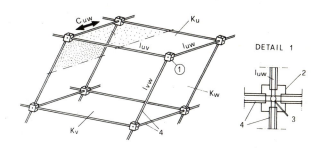

Fig. 29 Elements of similitude for
the three physical models.
(a) Elementary fractures
(b) Electrical resistance
(c) Circular pipe.

Fig. 30 Element of hydraulic analogue model
(1) Arrangement of conduits. Detail I:
Cross-section parallel to K_u
(2) Perspex prism
(3) Piezometric measurement point
(4) Perspex pipes.

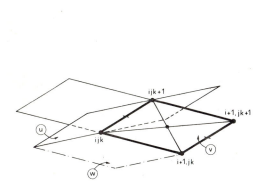

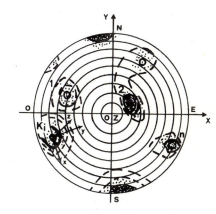

Fig. 31 Elements of the mathematical model with
elementary surfaces.

Fig. 32 Stereographic representation of fracturing in a rock mass with n̲ systems of
fractures.

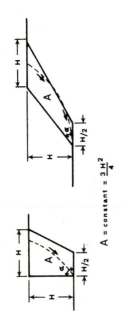

Fig. 34 Sliding curve and drainage criterion for different slope angles.

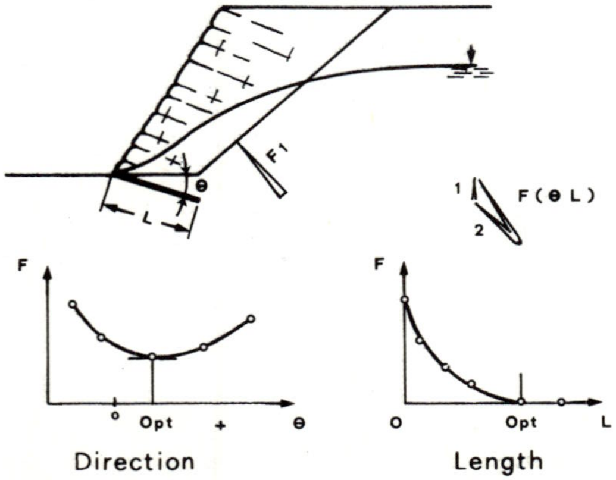

$$A = \text{constant} = \frac{3\,H^2}{4}$$

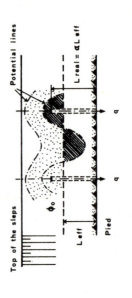

Fig. 36 Effective length of a parallel drainage network (plan view).

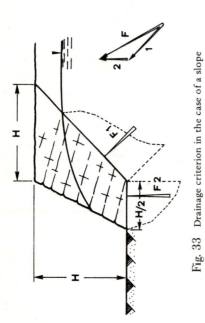

Fig. 33 Drainage criterion in the case of a slope

Fig. 35 Theoretical efficiency of a drainage system.

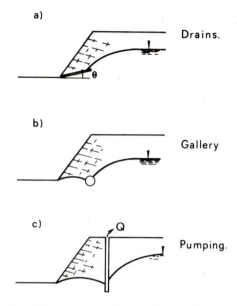

Fig. 37   Various possible ways for slope drainage.

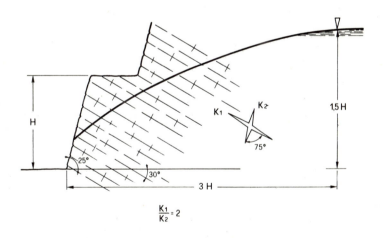

Fig. 38   Geometrical and hydraulic data of the example considered.

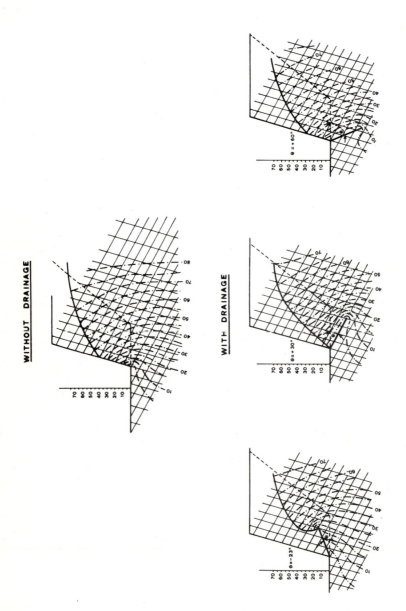

Fig. 39   Flow nets in a slope with different drain orientations (case of fig. 38a).

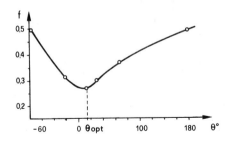

Fig. 40   Characteristic efficiency curve of drainage
(corresponding to fig. 35).

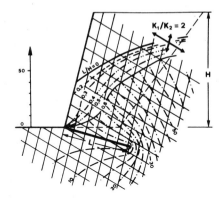

Fig. 41   Flow networks for various drain lengths.

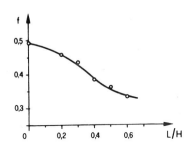

Fig. 42   A characteristic drainage curve for cases
outlined in figure 41.

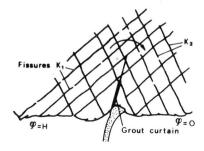

Fig. 43   Horizontal cross-section of the abutment
of the dam shown in fig. 25.

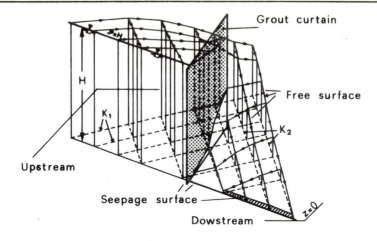

Fig. 44   Flow network in the dam abutment — case of fig. 43.

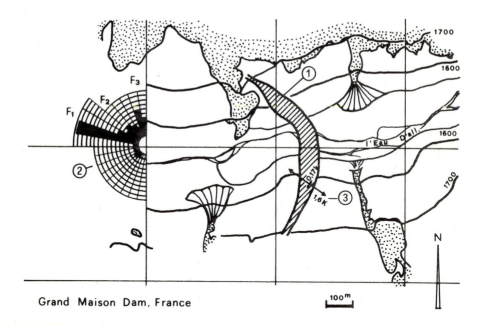

Grand Maison Dam, France

outcrops

Gneiss, Amphibolites   scree, alluvia

Fig. 45   Relative arrangement and principal structural elements
          (1) Dam
          (2) Fracture direction (subvertical)
          (3) Horizontal anisotropy of the permeability.

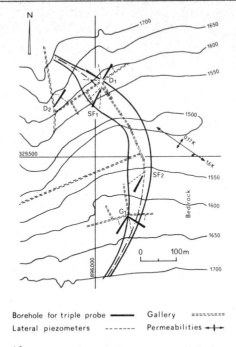

Borehole for triple probe ━━━━━   Gallery   ═══════
Lateral piezometers   - - - - - -   Permeabilities ◂╋▸

Fig. 46   Lay-out of reconnaissance means: Galleries and
boreholes.

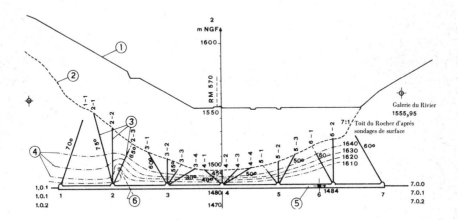

Fig. 47   Piezometric survey in the sub-fluvial gallery
Transverse cross-section
(1) Valley section (scree and alluvia)
(2) Roof of the bed-rock
(3) Boreholes
(4) Sub-fluvial gallery
(5) Potential lines
(6) Large fracture

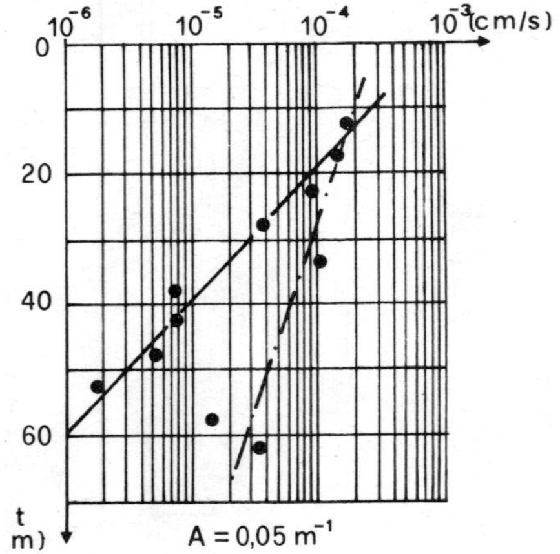

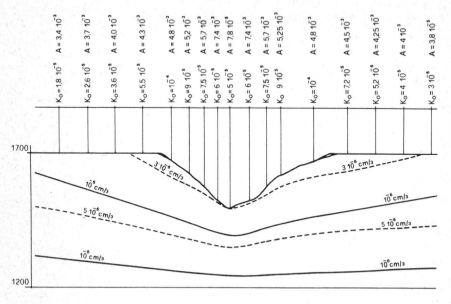

Fig. 49 Law of the distribution of the absolute rate of permeability in natural conditions. Transverse cross-section
K = Ko Exp (-At), K in cm/s, A in m^{-1}.

Fig. 48 The law of variation of the permeability according to the depth (Log Ko - Log K = At)

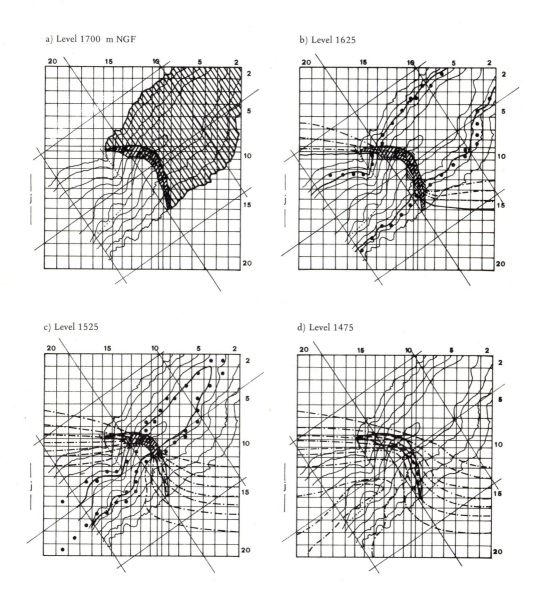

a) Level 1700 m NGF

b) Level 1625

c) Level 1525

d) Level 1475

Fig. 50 Results of simulation by three-dimensional mathematical model Horizontal cross-sections.

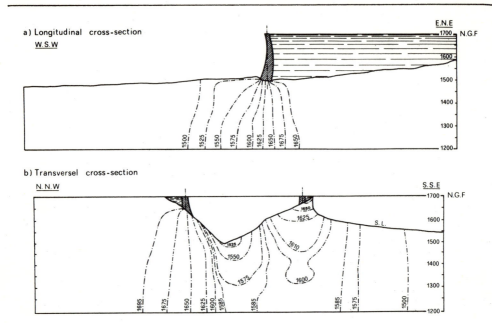

Fig. 51   Transverse and longitudinal cross-sections of the flow network (see fig. 50).

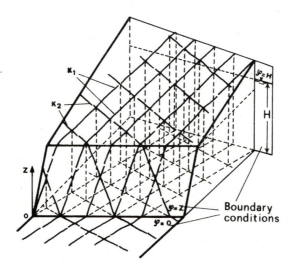

Fig. 52   Groundwater flow in slope in fractured rock.

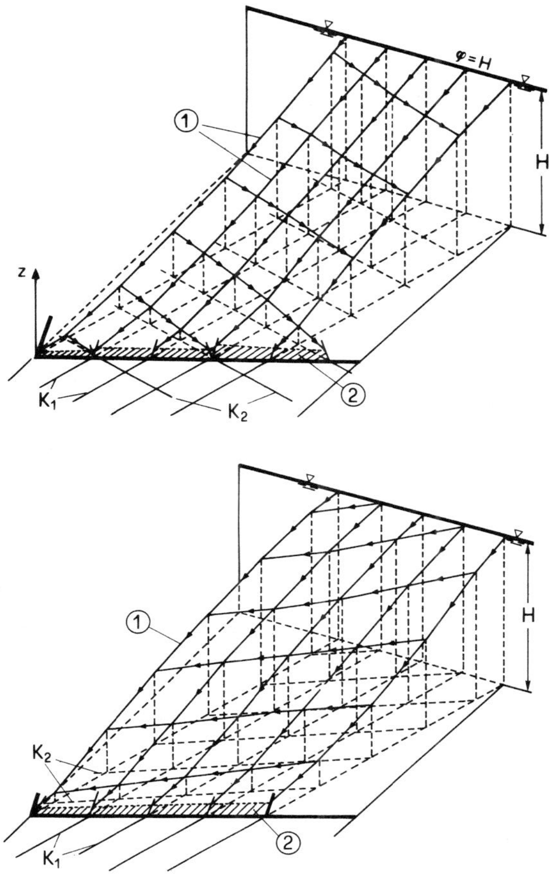

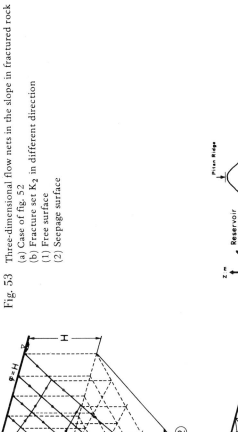

Fig. 53 Three-dimensional flow nets in the slope in fractured rock
(a) Case of fig. 52
(b) Fracture set K_2 in different direction
(1) Free surface
(2) Seepage surface

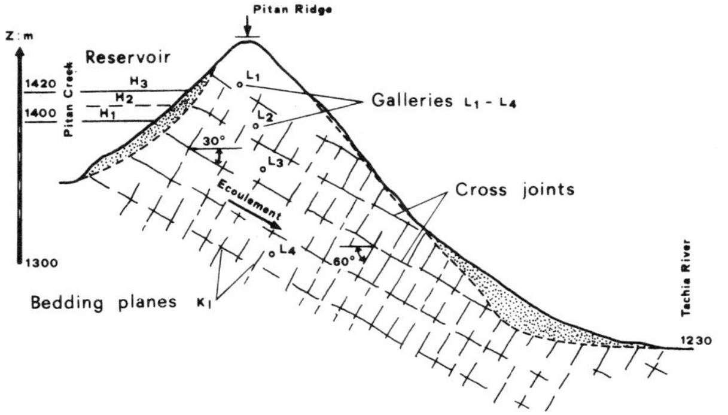

Fig. 54 Schematic cross-section of Pitan Ridge (Tachien project, Formosa) Hydraulic and geometric data.

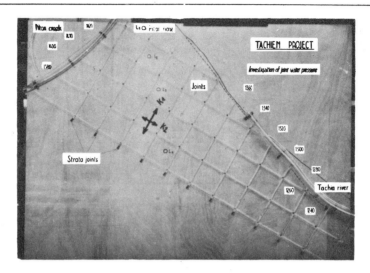

a

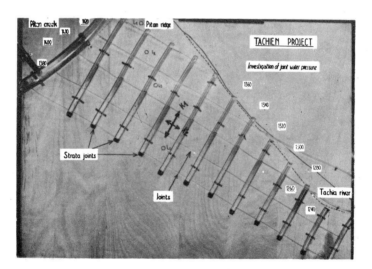

b

Fig. 55   Hydraulic models for the hydrogeotechnical study of the Pitan Ridge
         (a) Ratio of hydraulic conductivities: $K_1/K_2 = 2$
         (b) Ratio of hydraulic conductivities: $K_1 \gg K_2$.

a) Hypothese A

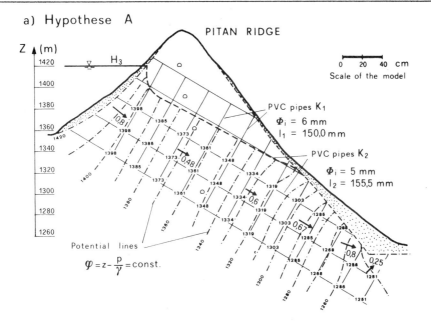

b) Hypothese B

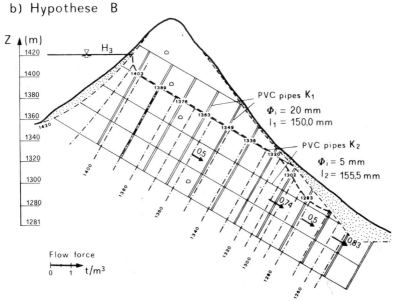

Fig. 56   Hydraulic models for the hydrogeotechnical study of the Pitan Ridge
            (a) Ratio of hydraulic conductivities: $K_1/K_2 = 2$
            (b) Ratio of hydraulic conductivities: $K_1 \gg K_2$.

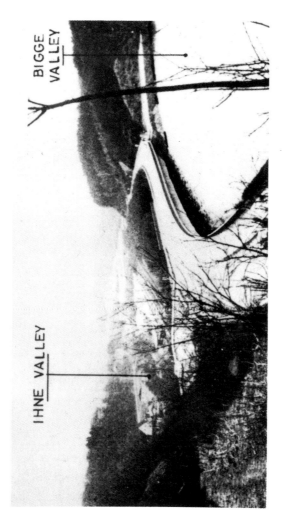

Fig. 57 View of the ridge "Kraghammer Sattel" at the Bigge damsite, West Germany.

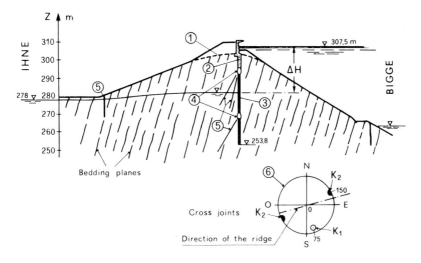

Fig. 58   Cross section of the ridge "Kraghammer Sattel"
(1) Natural profile   (4) Controlling galleries
(2) Concrete piling   (5) Piezometers and drainage
(3) Grout curtain   (6) Main fractures in zone I.

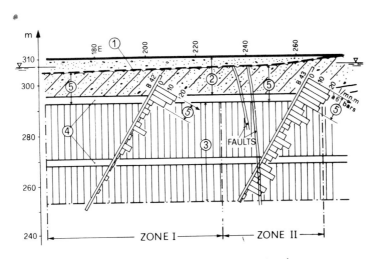

Fig. 59   Longitudinal cross-section of Kraghammer Sattel
(1) Natural surface   (4) Control galleries
(2) Concrete wall   (5) Decompressed zone
(3) Grout curtain   (6) Water tests

Cross section

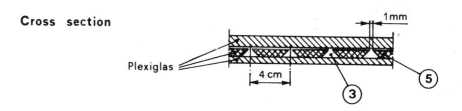

Fig. 60   Hydraulic model for study of flows in zone I of Kraghammer Sattel
        (1) Reservoir at 307.5 m level   (4) Downstream level 278 m
        (2) Grout curtain   (5) Simulation of cross- joints K_2
        (3) Simulation of the bedding

a) Zone I

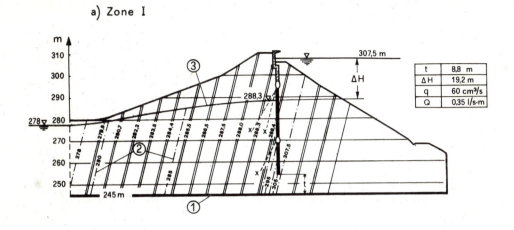

b) Zone II

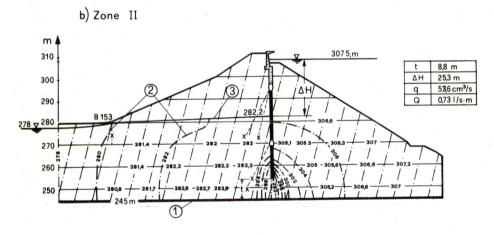

Fig. 61   Flow nets in the rock ridge "Kraghammer Sattel" (Two of 13 cases considered)
(1) Impervious boundary   x = Piezometers
(2) Equipotential lines   q = Flow rate measured on model
(3) Free surface   Q = Flow rate measured in situ.

CONTENTS